For a complete listing of titles in the
Artech House Radar Series,
turn to the back of this book.

Radar RF Circuit Design

Nickolas Kingsley

J. R. Guerci

ARTECH
HOUSE

BOSTON | LONDON
artechhouse.com

Library of Congress Cataloging-in-Publication Data
A catalog record for this book is available from the U.S. Library of Congress.

British Library Cataloguing in Publication Data
A catalogue record for this book is available from the British Library.

Cover design by John Gomes

ISBN 13: 978-1-60807-970-4

© 2016 ARTECH HOUSE
685 Canton Street
Norwood, MA 02062

10 9 8 7 6 5 4 3 2 1

For Nicole

Contents

Acknowledgments

We would like to thank our families for providing endless support in the creation of this work. Writing a book is a labor of love, and it is their love and inspiration that kept our fingers tapping. In addition, we offer many thanks to the reviewers for their insight, guidance, and encouragement throughout the writing process.

We would also like to thank Keysight Technologies for generously providing access to its analysis software for generating the simulation examples provided. Our industry has come a long way from vellum paper and slide rules thanks to trailblazers in electronic design like Keysight Technologies.

Finally, we'd like to thank our readers for their interest and passion in the subject. We sincerely believe that tighter collaboration and understanding between system designers and component designers will lead to advancements in radar technology like never before. It all begins with crossing the chasm from system to component level. Let's get started.

Part I
Microwave Background

1

Crossing the Chasm from System to Component Level

There is in practice a clear division between "system-level" and "component-level" thinking. Engineers typically classify themselves as system or component designers, and indeed that is the case for the authors of this book. Conferences also often differentiate between system and component focus. Funding sources [like government broad agency announcements (BAAs)] seek to make advancements at one level or the other. Consequently, it's easy to see why a chasm naturally exists. The chasm is so deep that even the vernacular is different. Table 1.1 lists various terms commonly used at the system and component level.

1.1 Basic Radar Systems Overview

To first order, the performance of a radio detection and ranging, or radar, system is driven by the size of the antennas employed and the power it can generate. In practice however, both these quantities are highly constrained due to a multitude of factors including size, weight, power, and cost (SWAP-C). Thus, real-world radar systems engineering quickly evolves into "getting the most out of what is available." In addition to clever signal-processing techniques, such as pulse compression and adaptive processing, optimizing component-level performance is key—and the subject of this book. This is made clear in the following description of a basic radar system.

At the simplest level, radar systems operate by sending out a signal and precisely measuring the reflection. The distance from the radar to the target is determined by the time required for the signal to return. The velocity of the target can be determined from the measured Doppler shift or simply by observing

Table 1.1
Common System- and Component-Level Terminology

Term	System Level	Component Level
Power	Effective radiated power (ERP)	Output power (P_{out})
Linearity	Adjacent channel power ratio (ACPR)	3rd-order intermodulation (IM3)
Efficiency	Prime power consumption	Power added efficiency (PAE)
Noise	Signal-to-noise ratio (SNR)	Noise figure (NF)
Waveform	Chirp	Two-tone

the change in range over time. Monostatic radars generally use one antenna (or two colocated) to perform both transmit and receive functions, whereas bistatic radars use two different antennas separated by a distance much greater than the wavelength of operation.

1.1.1 Radar Transmitters

A radar transmitter generates a frequency signal of interest, amplifies the signal to a power level sufficient to reach the desired maximum range of operation, and radiates that signal into the environment through an antenna. The signal will either come in contact with an object and scatter, or it will eventually propagate far enough that it is naturally absorbed into the atmosphere or proceed further to travel into space. Generally, radar transmitters are benchmarked against the following parameters:

- Power-aperture product, or the maximum operating range; a function of the ERP out of the transmit and the two-way antenna gain (transmit and receive antenna gain).

- Operating frequency, or how the signal behaves when propagating through the atmosphere; it is also related to the amount of signal bandwidth that is practical. Higher operating frequencies allow for wider bandwidth and, in turn, finer range resolution (which is proportional to the reciprocal of the operating bandwidth).

- Prime power, or the energy (i.e., fuel) required to operate the system, which is always greater than the ERP.

- Power output variation, or how the power fluctuates with frequency.

- Intermodulation, or the ability to generate signals as spectrally clean as possible.

- Robustness, or the ability for a system to survive or recover from a mechanical or electrical disturbance.

1.1.2 Radar Receivers

A radar receiver detects the scattered signal from the environment, amplifies the signal to a level suitable for processing, filters unwanted components (if possible), and extracts usable information. This information includes, for example, the size, position, and direction of objects within the operating range. Generally, radar receivers are benchmarked against the following characteristics:

- Sensitivity: The ability to detect signals above the background noise floor;
- Selectivity: The ability to differentiate signals from one another;
- Intermodulation: The ability to suppress the creation of additional non-linearities from the input signals;
- Frequency locking: The ability to lock on and track detected signals;
- Leaky radiation: The ability to prevent detected signals from leaking to the antenna and reradiating;
- Robustness: The ability to survive or recover from a mechanical or electrical disturbance.

Practical Note

One of the greatest advantages of radar is its ability to reach out over vast distances in all weather conditions. The Arecibo radio telescope in Puerto Rico has been used in radar mode to map the surface of Venus nearly 42 million kilometers (26 million miles) away! As one would expect, the signal reflection from a target many hundreds or even thousands of miles away can be many orders of magnitude less than that which was transmitted—as much as 100-dB attenuation or more.

1.1.3 Fundamental Equations

The basic principles of radar have been studied since the early 1900s. Although it is presented in many forms, the main radar equation is [1]:

$$P_r = \frac{P_t G_t G_r \lambda^2 \sigma}{(4\pi)^3 R^4} \qquad (1.1)$$

where P_r is the received power (watts), P_t is the transmit power (watts), G_t is the transmitter antenna gain (unitless), G_r is the receiver antenna gain (unitless), λ is the wavelength (meters, m), σ is the radar cross section (RCS) of the target (square meters, m^2), and R is the distance to the target (m).

Equation (1.1) is generally applied to the single pulse case, where the pulse width τ is inversely proportional to the receiver bandwidth B. If pulse compression is employed, whereby the pulse width is much larger but still retains the same bandwidth, a multiplicative term in the numerator of (1.1) is included, which is equal to the time-bandwidth product τB.

Additionally, if Doppler processing is employed, whereby a set of N pulses (compressed if pulse compression is employed) are fed to a Doppler filter bank, an additional gain term in the numerator of (1.1) proportional to N is added (coherent processing assumed).

The effective target signal strength is directly proportional to the received power level. Per the radar equation, smaller targets can be detected by doing the following:

- Increasing the transmit power level;
- Increasing the transmit and/or receive antenna gain (that is, making them more directive);
- Decreasing the distance from the target;
- Increasing the dwell time (i.e., illuminating the target for longer periods);
- Increasing the pulse width via pulse compression techniques;
- Optimizing the choice of polarization (target RCS often varies with polarization).

By far, the biggest impact can be gained by decreasing the distance to the target (the infamous "R to the fourth" relationship). Since this is generally not possible, it is necessary to optimize all of the above to achieve requisite performance.

If the target is in motion, the received signal will be shifted in frequency from the transmitted signal. This change in frequency is due to the Doppler shift and can be calculated by [1]:

$$f_r = f_o \pm \frac{2 v f_o}{c} \tag{1.2}$$

where f_r is the received frequency (Hz), f_o is the original (transmitted) frequency (Hz), v is the target velocity projected along the radial line-of-sight direction (meters per second, m/s), and c is the speed of light (m/s). Note that the sign is positive when the target is approaching and negative when the target is receding.

For example, a target traveling at Mach 1 (340.3 m/s) will have a Doppler shift of 22.7 KHz on a 10-GHz signal. This is a very small difference that must be accurately measured if radial velocity is a required radar output.

In the real world, any received electrical signals are always accompanied by unwanted background signals called *noise*. Though there can be many sources of such interference, some are fundamental and completely unavoidable. The most ubiquitous is so-called *thermal noise*, which is due to random molecular motions in the radar components due to temperature. The thermal noise power (P_n) in a radar receiver at 290K is calculated by:

$$P_n = kT_oFB \qquad (1.3)$$

where k is Boltzmann's constant (1.38×10^{-23} W·s/K), T_o is the standard temperature (290K), B is the instantaneous receiver bandwidth (Hz), and F is the noise figure of the receiver (unitless) and is highly dependent on component selection and design—the focus of this book.

To be an effective radar, the target signal must have a power level that can be distinguished from the thermal noise *floor*. To ensure this is the case, the ratio between the signal level and the noise level must be carefully analyzed. This SNR can be determined by combining the earlier equations [1]:

$$SNR = \frac{P_tG_tG_r\lambda^2\sigma}{(4\pi)^3 R^4 kT_oFB} \qquad (1.4)$$

While the radar is in *search* mode, the target's location is unknown. Therefore, the radar must search a greater volume of space. The following relationship is the search radar equation [1].

$$SNR_{search} = \frac{P_{avg}A_e\sigma T_{fs}}{4\pi k\Omega L_s T_o FR^4} \qquad (1.5)$$

where P_{avg} is the average transmitted power (W), A_e is the effective aperture size (m²), σ is the RCS of the target (m²), T_{fs} is the frame search time (seconds, s), k is Boltzmann's constant, Ω is the search angle solid angle (steradian), L_s is the system loss (unitless), T_o is the standard temperature (290K), F is the noise figure (unitless), and R is the distance to the target (m). Note that both system losses and noise figure are directly related to the choice of "components and circuits" employed by the radar.

While the radar is in *track* mode, the target location is known to a degree sufficient to focus resources. Therefore, the antenna can be pointed directly at the target. The following relationship is the track radar equation [1].

$$SNR_{track} = \frac{P_t G_t G_r \lambda^2 \sigma}{(4\pi)^3 R^4 k T_o FL_s B} \qquad (1.6)$$

where P_t is the transmit power (watts, W), G_t is the transmitter antenna gain (unitless), G_r is the receiver antenna gain (unitless), λ is the wavelength (m), σ is the radar cross section of the target (m²), and R is the distance to the target (m), k is Boltzmann's constant, T_o is the standard temperature (290K), L_s is the system loss (unitless), and B is the noise bandwidth of the receiver (Hz).

1.1.4 Requirements on Components

From the SNR equations for searching and tracking, the parameters that effect performance are evident. However, a radar's power, antenna, operating frequency, and bandwidth are often dictated by the application and host platform—and the target cross section is rarely under the control of the radar. Therefore, the only design variables left to the radar engineer to enhance performance is the minimization of so-called system losses and noise figure [and possibly antenna efficiency, especially for electronically scanned antennas (ESAs)].

The front-end components (i.e., antenna, RF filter, amplifier, and wiring/cabling between these components) all contribute to the system thermal noise, and thus great care must be given to minimize these effects. Both the architecture and component performance can have a big impact. For example, if the RF filter and amplifier can be placed in close proximity to the antenna (perhaps "integrated into" the antenna), then significant wiring/cable losses can be mitigated. However once an architecture is set, all focus is on maximizing component performance, such as the filters and amplifiers.

When requirements flow down from the system level, they generally fall into five categories:

- Electrical: Performance required to achieve the system needs;
- Physical: Size and form factor that fits in the space allotted;
- Cost: Price point needed to be marketable;
- Operating conditions: Thermal, lifetime, and other environmental factors;
- Manufacturability: Constraints posed by the production facility.

Unfortunately, all too often, these flow-downs are dictated without an understanding of what is possible at the component level. Sometimes requirements are set that exceed physics; sometimes requirements are set below what could be achieved because they are presumed impossible.

> **Practical Note**
>
> One of the next-generation wideband radars under development had a stringent frequency flatness requirement. This requirement was flowed down from the system designers to the submodule design teams. To compensate for the +6-dB/octave gain slope inherent in the antenna array, the designers incorporated a lossy equalizer circuit. Separately, the amplifier front-end design team incorporated a lossy equalizer circuit to compensate for the –6-dB/octave gain slope inherent in their active component. Had the system designers understood the complementary nature of these components, 6 dB more power would have been available (magnifying the receive power level by four times)!

1.2 Introduction to Microwave Components

Microwave components are the building blocks for microwave assemblies, modules, subsystems, and ultimately systems. Passive components manipulate electrical signals without adding energy. Active components do the same, but they draw on energy from an external source and can thus potentially add energy (such as amplification) to a signal.

1.2.1 Fundamental Equations

The most fundamental behaviors of electromagnetics are described by Maxwell's equations. Although compact and perhaps deceptively "simple," they are extremely powerful in what they represent. A deep understanding of their usage is not necessary to be a good microwave designer, but they are referenced throughout the book [2]. A description of each follows.

Gauss' Law for Electricity

An electric charge produces an electric field, and the flux of that field passing through a defined closed surface is proportional to the total charge within that surface, or:

$$\oint_S \vec{E} \circ \hat{n} \, da = \frac{q_{enc}}{\varepsilon_o} \tag{1.7}$$

where $\vec{E}$ is the electric field (Newtons/Coulomb), "$\circ \hat{n}$" indicates the portion of the field perpendicular to the surface, *da* is an increment of surface area (m²), q_{enc} is the amount of charge within the surface (Coulombs, C), and ε_o is the free-space permittivity.

The equation can also be represented in differential form:

$$\vec{\nabla} \circ \vec{E} = \frac{\rho}{\varepsilon_o} \tag{1.8}$$

where "$\vec{\nabla}\circ$" indicates the mathematical function for divergence , $\vec{E}$ is the electric field (Newtons/Coulomb), ρ is the charge density (Coulomb/m^2), and ε_o is the free-space permittivity. Divergence is the sum of the derivatives in each of the axes, or:

$$\vec{\nabla} \circ \vec{A} = \frac{\partial A_x}{\partial x} + \frac{\partial A_y}{\partial y} + \frac{\partial A_z}{\partial z} \tag{1.9}$$

Gauss' Law for Magnetism

The total magnetic flux through any closed surface is zero, or:

$$\oint_S \vec{B} \circ \hat{n} da = 0 \tag{1.10}$$

where $\vec{B}$ is the magnetic field (Teslas), "$\circ\hat{n}$" indicates the portion of the field perpendicular to the surface, and da is an increment of surface area (m^2).

The equation can also be represented in differential form:

$$\vec{\nabla} \circ \vec{B} = 0 \tag{1.11}$$

where "$\vec{\nabla}\circ$" indicates the mathematical function for divergence and $\vec{B}$ is the magnetic field (Teslas).

Faraday's Law

A changing magnetic flux through a surface induces a force in any path on that surface and a changing magnetic field induces a circulating electric field, or:

$$\oint_C \vec{E} \circ d\vec{l} = -\frac{d}{dt}\int_S \vec{B} \circ \hat{n} da \tag{1.12}$$

where $\vec{E}$ is the electric field (volts/meter), "$\circ d\vec{l}$"indicates the portion of the field parallel to segment dl, "d/dt" is the rate of change with time, and "$\int_S \vec{B} \circ \hat{n} da$" is the magnetic flux.

The equation can also be represented in differential form:

$$\vec{\nabla} \times \vec{E} = -\frac{\partial \vec{B}}{\partial t} \tag{1.13}$$

where "$\vec{\nabla} \times$" indicates the mathematical function for curl, $\vec{E}$ is the electric field (volts/meter), and "$\partial \vec{B} / \partial t$" is the rate of change of the magnetic field with time. Curl is the tendency of the field to rotate in one of the coordinate planes, or:

$$\vec{\nabla} \times \vec{A} = \left(\frac{\partial A_z}{\partial y} - \frac{\partial A_y}{\partial z} \right) \hat{i} + \left(\frac{\partial A_x}{\partial z} - \frac{\partial A_z}{\partial x} \right) \hat{j} + \left(\frac{\partial A_y}{\partial x} - \frac{\partial A_x}{\partial y} \right) \hat{k} \quad (1.14)$$

Ampère-Maxwell Law

An electric current or changing electric flux through a surface produces a circulating magnetic field, or:

$$\oint_C \vec{B} \circ d\vec{l} = \mu_o \left(I_{enc} + \varepsilon_o \frac{d}{dt} \int_S \vec{E} \circ \hat{n} da \right) \quad (1.15)$$

where $\vec{B}$ is the magnetic field (Teslas), "$\circ d\vec{l}$" indicates the portion of the field parallel to segment dl, μ_o is the free space permeability, I_{enc} is the electric current (amps), ε_o is the free-space permittivity, "d/dt" is the rate of change with time, and "$\vec{E} \circ \hat{n} da$" is the electric flux through the surface.

The equation can also be represented in differential form:

$$\vec{\nabla} \times \vec{B} = \mu_o \left(\vec{J} + \varepsilon_o \frac{\partial \vec{E}}{\partial t} \right) \quad (1.16)$$

where "$\vec{\nabla} \times$" indicates the mathematical function for curl, $\vec{B}$ is the magnetic field (teslas), μ_o is the free-space permeability, $\vec{J}$ is the current density (amps/m²), ε_o is the free-space permittivity, and "$\partial \vec{E} / \partial t$" is the rate of change of the electric field with time.

1.2.2 Essential Components

Just as craftspeople must be familiar with their tools, microwave designers must understand the diversity of components at their disposal. Table 1.2 lists the most common RF components in alphabetical order with a brief explanation. Design strategies for most of these, or portions thereof, are discussed in this book.

1.3   Traveling Wave Tubes Versus Solid State

Radars used during World War II used magnetrons to create traveling wave tube amplifiers (TWTAs), pronounced "tweet-ahs". Today, TWTAs are still the primary technology for extremely high-frequency/high-power applications. Although transistors have paved the way for smaller, more affordable, easier to integrate, and more robust systems, there are certain power levels that still elude them. Klystrons and TWTAs that provide 10s of megawatts of power are available, whereas kilowatt solid-state amplifier modules are still in their infancy [3].

The focus of this book is on solid-state solutions since that is arguably the future of radar systems. Also, extremely high-power radars can be created from solid-state amplifiers if multiple coherent antennas are used, each driven by separate solid-state amplifiers. These are referred to as *active electronically scanned antennas* (AESAs). The ease of integration into AESA configurations, which offer reduced prime power requirements and versatility, makes them an ideal component in radar systems. Throughout the text, a comparison to TWTAs will be made for reference.

1.4   "How" Components are Connected Matters

Consider the two receiver configurations shown in Figure 1.1.

Figure 1.1(a) shows an antenna connected to an LNA via a conveniently sized run of coax cable. The reason this configuration is convenient is because the amplifier can be placed in a suitable location isolated from the elements, other RF interference, and perhaps with easy access to repairs. Lots of benefits! Now consider Figure 1.1(b), where the LNA is place directly behind the antenna. For the same antenna and LNA, which configuration has the lowest noise figure? Even without intimate RF design knowledge most engineers can easily reason that Figure 1.1(b) has a better noise figure since the coax cable, as an electrical component, adds loss and thermal noise. In fact, the added loss is proportional to the square root of frequency, the type of metal and dielectric used in the coax, and the diameter of the coax. Not something to be chosen casually!

The examples in Figure 1.1 illustrate that even the most basic aspects of combining RF components have a profound impact on overall performance. Despite the noise figure advantage of the Figure 1.1(b) configuration, there are many applications where it simply is not possible to colocate the LNA and antenna. Thus the goal in such situations is to minimize the distance to the greatest extent possible and, of course, to use the highest quality coax possible.

Other major component connection/configuration considerations include location of components to allow for waste heat removal (a major consideration in high-power applications), placement to minimize RF leakage into other components, and placement and packaging to meet often stringent

Table 1.2
Common RF Components with Descriptions and Symbols

Component	Description	Symbol
Absorber	Material (generally ferrite-loaded) that absorbs electromagnetic energy	
Antenna	Radiates electromagnetic energy into free space	
Attenuator	Resistive element that adds loss	
Balun	Transformer that converts a balanced signal (two signals with no fixed ground) to an unbalanced signal (clear common ground)	
Bias tee	Three-port network that combines DC with RF or takes a signal and splits it into DC and RF components	
Capacitor	Basic element that blocks DC and passes AC and that can also store charge	
Circulator	Three- (or more)port network that restricts the flow of electromagnetic energy to one direction	
Coupler	Four-port network that splits the input to two equal or unequal amplitude outputs and that has an isolation port	
Diode	Basic element that only passes current in one direction (the direction the triangle points)	
Diplexer	Three-port network that splits into two ports with different frequency responses	
Duplexer	Allows a transmitter and receiver to share a single antenna	
Equalizer	Flattens a response (such as gain) over frequency	
Filter	Changes the amplitude of a signal based on the frequency response (band pass filter shown)	
Inductor	Basic element that blocks AC and passes DC and stores magnetic flux	
Isolator	Two-port network that restricts the flow of electromagnetic energy to one direction	
Limiter	Prevents output power from exceeding a threshold	

Table 1.2   (continued)

Component	Description	Symbol
Low-noise amplifier (LNA)	Amplifier optimized for high gain and low noise generation	
Mixer (downconverter)	Multiplies an input signal (RF) by a fixed frequency (LO) to downconvert to an intermediate frequency (IF)	RF — ⊗ — IF, LO
Mixer (upconverter)	Multiplies an input signal (IF) by a fixed frequency (LO) to upconvert to an RF frequency (RF)	IF — ⊗ — RF, LO
Phase shifter	Modifies the phase response of a signal	
Power amplifier	Amplifier optimized for high output power	
Power combiner	Multiport network that combines multiple input ports into a single output port with increased amplitude	
Power splitter	Multiport network that splits a single input into multiple output ports with reduced amplitude	
Resistor	Basic element that attenuates voltage	
Switch	Basic element that directs a signal from one path to another	
Thermistor	Resistor with predictable temperature response	T
Transistor	Voltage-controlled resistor and basic element in an amplifier	
Varactor	Voltage or mechanically tunable capacitor	

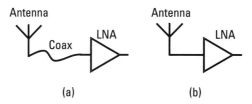

(a)   (b)

Figure 1.1   Two different configurations of identical receive antenna and LNAs: (a) LNA connected via coax cable and (b) LNA directly behind the antenna.

environmental operating conditions. These considerations will be discussed in this book.

Exercises

1. A radar detects a 10.2-KHz Doppler shift at 8.5 GHz on an approaching target. How fast is the target moving?

2. All magnets have a north pole and a south pole. Using Maxwell's Equations, explain why these poles cannot exist separately. (Hint: What is the significance of Gauss' Law for Magnetism being equal to zero?)

References

[1]   Richards, M., J. Scheer, and W. Holm, *Principals of Modern Radar—Basic Principals,* Edison, NJ: SciTech Publishing, 2010.

[2]   Fleisch, D., *A Student's Guide to Maxwell's Equations,* New York, NY: Cambridge University Press, 2008.

[3]   Chang, K., I. Bahl, and V. Nair, *RF and Microwave Circuit and Component Design for Wireless Systems,* New York, NY: Wiley, 2002.

Selected Bibliography

Belov, L., S. Smolskiy, and V. Kochemasov, *Handbook of RF, Microwave, and Millimeter-Wave Components,* Norwood, MA: Artech House, 2012.

Saad, T., *Microwave Engineers' Handbook,* Norwood, MA: Artech House, 1972.

Kinayman, N., and M. Aksun, *Modern Microwave Circuits,* Norwood, MA: Artech House, 2005.

Vizmuller, P., *RF Design Guide: Systems, Circuits, and Equations,* Norwood, MA: Artech House, 1995.

Maas, S., *The RF and Microwave Circuit Design Handbook,* Norwood, MA: Artech House, 1998.

Losee, F., *RF Systems, Components, and Circuits Handbook,* Norwood, MA: Artech House, 2005.

2

Introduction to Microwave Design

Microwave design starts by understanding wave theory—or the rules that govern how signals (energy) propagate through a material. Once that is understood, ways to manipulate those signals through amplification, attenuation, transformation, and filtering can be explored.

Chapter 2 begins by treating all microwave components like a black box with at least one external interface (or port). Higher-level functions, including those mentioned earlier, are implemented by connecting multiple components together at their ports. Just like trying to connect a square peg to a round hole, integrating microwave components requires handling the interface properly to avoid mismatch. Otherwise, performance will be reduced, and potentially the integrated circuit will not function at all.

This chapter discusses methods for describing a component's behavior by quantifying its port performance are discussed, along with the process for determining the effect of mismatch. In addition, the chapter explains how waves propagate through various microwave structures, materials, and discontinuities. These are the prerequisites for microwave design.

Throughout the text, the unit of length will be presented in the most appropriate form. For example, equations are usually expressed in terms of meters (m), but printed circuit board thickness and metal widths are usually published in mils or thousandths of an inch (in). To convert between units, use the following:

- 1 in = 1000 mils = 25.4 millimeters (mm);
- 1 mil = 25.4 microns (μm) = 0.0254 mm;
- 1 mm = 1000 μm = 39.37 mils = 0.03937 in.

2.1 Scattering Matrix

Radar system development begins as a block diagram that shows how the desired functionality is implemented, along with a set of performance and SWAP-C requirements. Some of those blocks include microwave components, such as filters, couplers, amplifiers, and antennas. When operating as a linear system (output is proportional to input), the scattering matrix, more commonly known as the *S-parameter*, is the preferred method for representing behavior at component ports (i.e., inputs and outputs). More specifically, it is a mathematical representation of the incident, reflected, and transmitted port voltages. As an example, for a two-port network (Figure 2.1), the scattering matrix is a ratio of the following:

- S_{11}: Reflected voltage amplitude at port 1 to incident voltage at port 1;
- S_{22}: Reflected voltage amplitude at port 2 to incident voltage at port 2;
- S_{21}: Voltage amplitude exiting port 2 to voltage incident at port 1;
- S_{12}: Voltage amplitude exiting port 1 to voltage incident at port 2.

Practical Note

S-parameter matrix notation (i.e., S_{11}, S_{21}, and S_{54}) is a concatenation of individual ports. Therefore, they should be read as such. For example, S_{11} should be read "S one one" not "S eleven."

Figure 2.1 provides an example of a two-port network. This same nomenclature can be applied to any linear *N*-port network.

S-parameters are commonly represented in log scale in decibels. (They can also be found in magnitude-phase formats.) Generally S_{11} is in negative decibels to indicate less power is reflected than is incident. S_{21} is negative for passive circuits (indicating loss) and positive for active circuits (indicating gain).

The return loss at port *m* in decibels can be calculated from

$$RL = 20\log\left|\frac{1}{S_{mm}}\right| \tag{2.1}$$

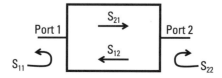

Figure 2.1 Example two-port network.

The insertion loss between ports m and n in decibels can be calculated from:

$$IL = 20\log\left|\frac{1}{S_{mn}}\right| \qquad (2.2)$$

Practical Note

Insertion loss is also called *attenuation*. Although decibels are the most widely used unit, sometimes nepers (*Np*) are also used. They can be calculated by:

$$1Np = \frac{20}{\ln 10}dB \approx 8.68589 dB \qquad (2.3)$$

$$1dB = \frac{1}{20\log e}Np \approx 0.11513 Np \qquad (2.4)$$

Return loss and insertion loss are always positive numbers (in decibels). Negative loss implies positive gain which may not be the case. This is not to be confused with S_{11} (the S-parameter representation of return loss) or S_{21} (the S-parameter representation of insertion loss), which can be negative in decibels. For example, a 10-dB return loss is the equivalent of -10 dB S_{11}. In circuit design, it is preferable for S_{11} to always be negative.

For most applications, a 10-dB return loss is considered sufficient for proper operation. This computes to 0.1 or 10% reflected power. Other applications require 20 dB, or 1% reflection. When analyzing a chain of components, it is important to take return loss into consideration. This is especially important in radars where often good return loss must be sacrificed to achieve other SWAP-C considerations.

Some properties of S-parameters [1] are described as follows:

- Any perfectly matched port has $S_{mm} = 0$ (no signal is reflected);
- Reciprocal networks have $S_{mn} = S_{nm}$, which is true for any passive linear network;
- Passive circuits have $|S_{mn}| \leq 1$;
- Active linear circuits can have $|S_{mn}| > 1$.

2.2 Matching Networks

Most systems have standardized on 50+j0Ω system impedance (50Ω marks a compromise between 30Ω, which is best for power handling, and 77Ω, which is best for minimizing loss, in a coax line). For ease of integration, all lower-level assemblies and components conform to this impedance. Achieving perfect 50Ω operation is impractical, but it is the component designer's mission to get as close as possible while meeting all other criteria.

Matching networks attempt to bridge the gap between inherently low-impedance circuits (i.e., high-power amplifiers), high-impedance circuits (i.e., antennas), and the 50Ω system. In order to minimize the impact on system efficiency, matching networks must be as low-loss as possible. They must also support the complete bandwidth, power-handling, and phase-stability demands on the components they connect. A properly matched receiver will provide significantly better SNR than a poorly matched receiver.

In a perfect system, components in a block diagram would be cascaded together, and the total loss of the circuit would be equivalent to the sum of the individual component losses. However, in a practical system, not all components will provide a perfect system impedance at the ports. Between any two connected ports, some of the energy will be reflected back (reflected power) instead of delivered to the next component (incident power). To minimize reflections and maximize transmission, matching networks can be added to convert one impedance to another.

During the development phase of a program, incorporating tunable matching networks can give designers the ability to compensate for circuit-to-circuit variation. This is especially useful when commercial off-the-shelf (COTS) components are used and cannot be modified.

2.2.1 Quantifying Mismatch

A transmission line of length l and characteristic impedance Z_o is terminated by a load Z_L as shown in Figure 2.2. The impedance looking into the circuit is Z_{in}, and the ratio of reflected power (reflection coefficient) is Γ_L.

If $Z_L = Z_o$ and the transmission line is lossless, there is no impedance mismatch, and Γ_L is zero regardless of the transmission line length. In practical systems, this is never the case (at least at all frequencies). We can quantify the situation by:

$$\Gamma_L = \frac{Z_L - Z_O}{Z_L + Z_O} \tag{2.5}$$

Alternatively, that equation can be rearranged to solve for Z_L.

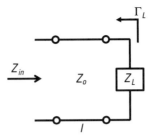

Figure 2.2 Transmission line terminated with load Z_L.

$$Z_L = Z_O \frac{1 + \Gamma_L}{1 - \Gamma_L} \tag{2.6}$$

Figure 2.3 plots Z_L versus Γ_L for a Z_o of 50Ω. If Γ_L is 0, there is no reflection and Z_L is 50Ω. If Γ_L is –1, Z_L is 0Ω (short circuit). If Γ_L is +1, Z_L is ∞Ω (open circuit). Even a small impedance mismatch can cause significant reflected power.

The impedance looking into the transmission line and load is a function of Z_o, Z_L, l, and the wavelength (λ) [2].

$$Z_{in} = Z_O \frac{Z_L + jZ_O \tan\left(\frac{2\pi l}{\lambda}\right)}{Z_O + jZ_L \tan\left(\frac{2\pi l}{\lambda}\right)} \tag{2.7}$$

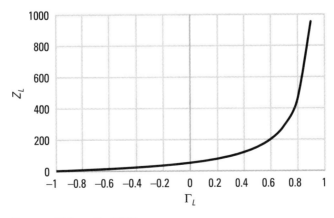

Figure 2.3 Z_L versus Γ_L for a Z_o of 50Ω.

For lossy transmission lines, the guided wavelength (λg, discussed later) should replace λ.

Practical Note

From (2.7), we can determine the impedance looking into a short ($Z_L = 0\Omega$) and open ($Z_L = \infty\Omega$) load. The short circuit input impedance is $jZ_o \tan(2\pi l/\lambda)$, and the open circuit impedance is $-jZ_o \cot(2\pi l/\lambda)$.

If an incident voltage (V^+) encounters the transmission line circuit shown in Figure 2.2, the incident power (P_{in}, W) can be determined by:

$$P_{in} = \frac{\left|V^+\right|^2}{Z_O} \tag{2.8}$$

If there is a load mismatch (Γ_L), the reflected voltage (V^-) and reflected power (P_r, W) can be calculated by:

$$P_r = \frac{\left|V^-\right|^2}{Z_O} = \frac{\left|V^+\right|^2 \left|\Gamma_L\right|^2}{Z_O} = P_{in}\left|\Gamma_L\right|^2 \tag{2.9}$$

If the input power (P_{in}) and reflected power (P_r) are known, then the transmitted power (P_t, W) can be calculated by:

$$P_t = P_{in} - P_r = P_{in}\left(1 - \left|\Gamma_L\right|^2\right) \tag{2.10}$$

Additionally, the return loss in decibels can be calculated by:

$$RL = -10\log\frac{P_r}{P_{in}} = -20\log\left|\Gamma_L\right| \tag{2.11}$$

$$\Gamma_L = 10^{\frac{-RL}{20}} \tag{2.12}$$

Another parameter for quantifying how well a load is matched is the voltage standing wave ratio (VSWR) (pronounced "viz-warr"). The value is written as a ratio compared to 1. A perfect match would have a VSWR of 1:1. A commonly specified VSWR is 2:1, which corresponds to a return loss of 9.54 dB and a reflection coefficient of 0.333. These values can be calculated from (where R_L is in decibels):

$$VSWR = \frac{|V_{max}|}{|V_{min}|} = \frac{1+|\Gamma_L|}{1-|\Gamma_L|} = \frac{1+10^{\frac{-RL}{20}}}{1-10^{\frac{-RL}{20}}} \qquad (2.13)$$

$$RL = -10\log\left[\left(\frac{VSWR-1}{VSWR+1}\right)^2\right] \qquad (2.14)$$

The mismatch loss in decibels due to a reflection can be calculated by:

$$ML = -10\log\left(1-\Gamma^2\right) = -10\log\left[1-\left(\frac{VSWR-1}{VSWR+1}\right)^2\right]$$
$$= -10\log\left[1-\left(10^{\frac{-RL}{20}}\right)^2\right] \qquad (2.15)$$

Practical Note

One of the challenges with specifying VSWR is that the quantity is scalar (only amplitude, no phase, information is presented). Therefore, both an open circuit and a short circuit have the same VSWR (∞:1). When converting from VSWR, Γ_L must be represented as the absolute value.

$$|\Gamma_L| = \frac{VSWR-1}{VSWR+1} \qquad (2.16)$$

Table 2.1 lists the return loss, mismatch loss, VSWR, and percent power transmitted versus Γ_L.

2.2.2 Graphically-Based Circuits

Part II will discuss how to determine the ideal impedances for a component. For an amplifier, for example, these would be the optimal source and load impedances for power, efficiency, and linearity, among other factors. This section assumes that these impedances are known. One of the simplest, most visual, and most intuitive methods for designing a matching network is to use a Smith

Table 2.1
Return Loss, Mismatch Loss, VSWR, and Percent Power
Transmitted Versus Γ_L

Γ_L	RL (decibels)	ML (decibels)	VSWR	Percent Transmitted
−1	0.00	∞	∞	0
−0.9	0.92	7.21	19.00	19
−0.8	1.94	4.44	9.00	36
−0.7	3.10	2.92	5.67	51
−0.6	4.44	1.94	4.00	64
−0.5	6.02	1.25	3.00	75
−0.4	7.96	0.76	2.33	84
−0.3	10.46	0.41	1.86	91
−0.2	13.98	0.18	1.50	96
−0.1	20.00	0.04	1.22	99
0	∞	0.00	1.00	100
0.1	20.00	0.04	1.22	99
0.2	13.98	0.18	1.50	96
0.3	10.46	0.41	1.86	91
0.4	7.96	0.76	2.33	84
0.5	6.02	1.25	3.00	75
0.6	4.44	1.94	4.00	64
0.7	3.10	2.92	5.67	51
0.8	1.94	4.44	9.00	36
0.9	0.92	7.21	19.00	19
1	0.00	∞	∞	0

chart. A Smith chart is a graphical representation of the plane of real (*Re*) and imaginary (*Im*) impedances. Figure 2.4 shows a Smith chart.

To be as versatile as possible, all impedance values are normalized to the system impedance:

$$Z_n = \frac{Z}{Z_o} \qquad (2.17)$$

where Z_n is the impedance plotted (Ω), Z is the impedance of interest (Ω), and Z_o is the system impedance (Ω). In most cases, Z_o is 50Ω so all impedances are divided by 50 before plotting. The center of the chart marks the system impedance and is always 1. The leftmost point of the chart marks a short circuit, or Z_n = 0. The rightmost point marks an open circuit, or $Z_n = \infty$. That line along the equator, called the *real line*, includes every normalized resistance from 0 to ∞.

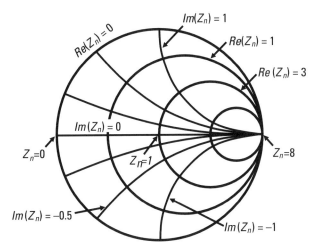

Figure 2.4   Classic Smith chart.

When the imaginary part of the impedance (called the *reactance*) is positive, the impedance lies in the northern hemisphere $(Im[Z_n] > 0)$. When the imaginary part of the reactance is negative, the impedance lies in the southern hemisphere $(Im[Z_n] < 0)$.

Plotting (and reading) an impedance from the chart requires three steps. Figure 2.5 shows how a normalized impedance of $0.40+j0.60\Omega$ is plotted. First, the circle connected to the real part of the impedance along the real line is found. In Figure 2.5, a circle that connects $Re(Z_n) = 0.40$ is drawn. Second, the radial line that connects to the imaginary part of the impedance is found. The $Im(Z_n) = 0.60$ arc is drawn in Figure 2.5. Third, the intersection of the circle and the arc is the normalized impedance.

Five normalized impedances are plotted in Figure 2.6.

There are many uses of the Smith chart, but the most popular are listed as follows:

- To transpose between Γ_L and Z_L;
- To transpose between Z_{in} and Z_L;
- To transpose between impedance (Z) and admittance $(Y = 1/Z)$;
- To determine VSWR;
- To perform impedance matches.

A Smith chart is probably most widely used for impedance matching because it is a visual (and intuitive) way to design a circuit. Lumped elements move around the Smith chart as shown in Figure 2.7. Series elements are placed in

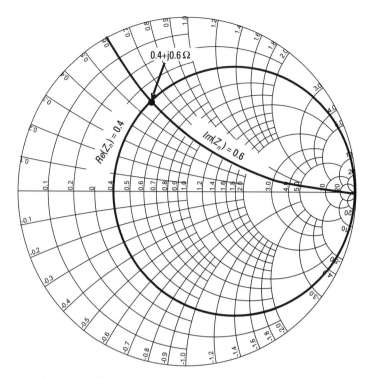

Figure 2.5   Plot of 0.40+j0.60Ω on a Smith chart.

line (or parallel) with the main signal path. Shunt elements are placed perpendicular with the main signal (often leading to ground). Series reactive elements move along constant resistance circles on the impedance Smith chart since the resistance is unchanged. Shunt reactive elements move along the constant conductance circles (the inverse of a resistance circle) since the conductance is unchanged. Along those same lines, series resistors (which have no reactive part) move along the constant reactance curves, and shunt resistors move along the constant susceptance curves [2].

Along the border of a Smith chart is a ruler (omitted from Figure 2.7) denoting the magnitude of the imaginary axis (the reactance or the conductance). To determine how far a component will rotate around the imaginary axis, the equivalent impedance or admittance is calculated. The total circumference of the Smith chart is equivalent to a half-wavelength.

The impedance of a series inductor, capacitor, and resistor can be calculated by:

$$Z_{series,L} = j(2\pi f)L \qquad (2.18)$$

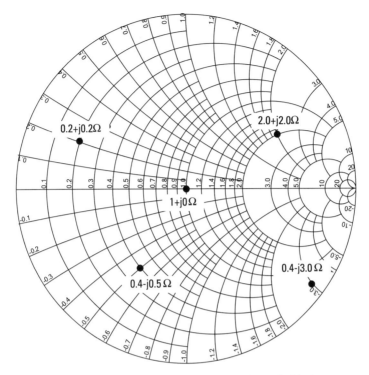

Figure 2.6 Five random normalized impedances are plotted on a Smith chart.

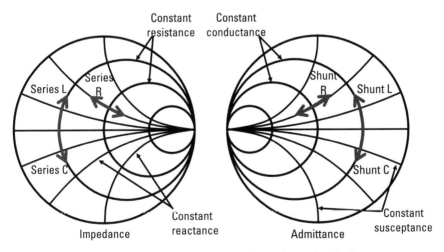

Figure 2.7 Series and shunt resistor, inductor, and capacitor on Smith chart.

$$Z_{series,C} = -j\frac{1}{(2\pi f)C} \tag{2.19}$$

$$Z_{series,R} = R \qquad\qquad (2.20)$$

The impedance of a shunt inductor, capacitor, and resistor can be calculated by:

$$Y_{shunt,L} = -j\frac{1}{(2\pi f)L} \qquad\qquad (2.21)$$

$$Y_{shunt,C} = j(2\pi f)C \qquad\qquad (2.22)$$

$$Y_{shunt,R} = \frac{1}{R} \qquad\qquad (2.23)$$

Notice that for all elements except resistors, the magnitude of the impedance shift is frequency-dependent. Let's match 50Ω to the example impedance presented earlier ($Z_n = 0.40+j0.60\Omega$) at 10 GHz with a system impedance 50Ω. There are an infinite number of ways to design a matching network. Let's first use a shunt capacitor to slide on the admittance chart along the constant conductance circle until we reach the constant $Z = 1$ resistance circle. Then, we'll use a series capacitor to slide on the impedance chart along the constant resistance circle to the origin.

The steps, shown in Figure 2.8, are described as follows. At $Z_n = 0.40+j0.60\Omega$, we can see on the admittance chart, this is on the −j1.15 susceptance curve. The point we want to slide to is on the −j0.42 susceptance curve. This delta is j0.73. Since we're dealing with normalized impedance/admittance, we must divide by Z_o (50) to get $Y_{shunt,C} = j0.015$. Solving for C using (2.22) gives 0.23 pF.

From our current position, we are on the j0.55 reactance curve. We want to slide to the real line, so the delta is j0.55. Once we normalize to Z_o, we get $Z_{series,C} = 27.5$. Solving for C using (2.19) gives 0.58 pF.

The final solution is a shunt 0.23 pF capacitor followed by a series 0.58 pF capacitor to 50Ω.

2.2.3 Distributed Matching Networks

In addition to lumped-element matching networks, transmission-line based (or distributed) elements can also be used. Any two resistances can be matched using a quarter-wave transformer. Matching R_L to Z_o, for example, would require a transmission line of resistance $\sqrt{R_L Z_o}$ and length $\lambda/4$. Matching to a complex

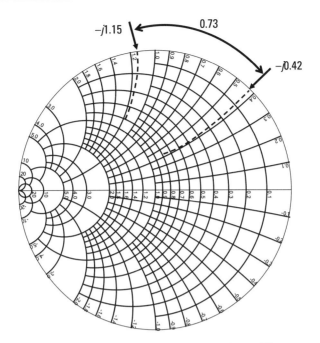

Figure 2.8   Plotting $0.40+j0.60\Omega$ with the equivalent of 0.23 pF at 10 GHz.

load can be accomplished by first using a single stub tuner to remove the reactive part, then using a quarter-wave transformer to match the resistive part.

Since quarter-wave transformers are inherently narrowband, a multisection network can be used for improved bandwidth as shown in Figure 2.11.

The input impedance is determined by:

$$Z_{in} = \left(\frac{Z_1 \cdot Z_3 \cdot Z_5 \cdot ...}{Z_2 \cdot Z_4 \cdot Z_6 \cdot ...} \right)^2 Z_l^{-1^N} \tag{2.24}$$

A three-section quarter-wave transformer can be used to match 5Ω to 50Ω using resistances 8.5Ω, 20Ω, and 37Ω as shown in Figure 2.12. This topic will be discussed further in Chapter 4.

2.3   Methods of Propagation

Before we can learn how to design components that manipulate electromagnetic energy (i.e., amplify, attenuate, filter, and radiate), we must first understand how waves propagate. Electromagnetic theory explains how waves travel through materials and structures. There are common techniques for directing electromagnetic energy from one place to another. The background and equa-

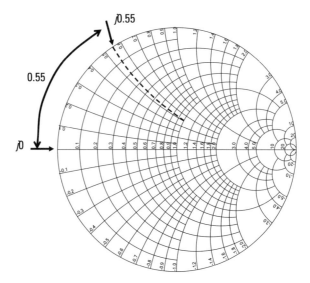

Figure 2.9   Adding a series 0.58 pF to Figure 2.8 at 10 GHz.

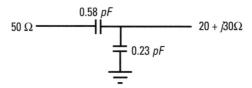

Figure 2.10   Final circuit to match 50Ω to 20+j30Ω.

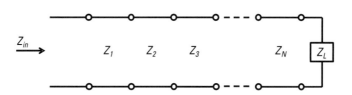

Figure 2.11   Multisection quarter-wave transformer.

tions will be explained in this chapter and used in the design examples in all subsequent chapters.

2.3.1   Wave Modes

Electromagnetic signals travel in waves, and there are a few different types that can propagate in waveguides. The type that is desired for most applications is the transverse electromagnetic (TEM) wave. TEM waves can propagate when-

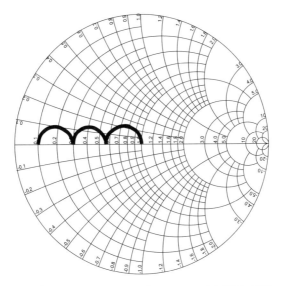

Figure 2.12   Three-section quarter-wave transformer to match 5Ω to 50Ω.

ever there are two or more conductors. A closed conductive waveguide, for example, would not support TEM propagation. Higher-order modes, including transverse electric (TE) and transverse magnetic (TM), can propagate through any single-conductor or multiconductor structure.

A TE mode is present whenever the direction of propagation is perpendicular to the electric field and parallel to the magnetic field. The TE mode is further subdivided into submodes (TE_{mn}), where m is the number of half wavelengths across the broad dimension of the electric field and n is the number of half wavelengths across the narrow dimension. m or n can be zero if their respective dimension is less than a half-wavelength long.

A TM mode is present whenever the direction of propagation is perpendicular to the magnetic field and parallel to the electric field. As with TE, the TM mode is further subdivided into submodes (TM_{mn}), where m is the number of half wavelengths across the broad dimension of the magnetic field and n is the number of half wavelengths across the narrow dimension. m or n can be zero if their respective dimension is less than a half-wavelength long.

Although TEM propagates across all frequencies, higher-order modes only propagate at frequencies above a certain threshold, known as the *cutoff frequency*. Preventing the excitation of TE and TM modes is important because those signals will create mixing products with the TEM signal through multiplication. This will certainly degrade linearity (or spectral purity) and can even cause instability or catastrophic failure in active components. Ways to prevent higher-order mode excitation will be addressed in Part II.

2.3.2 Coaxial Cables (Coax)

The coax cable is probably the most familiar structure, because it is used to bring cable television into residences. Coax lines have a center conductor surrounded by a grounded shield. In Figure 2.13, d is the center conductor diameter, D is the inner diameter of the ground shield, and ε_r/μ_r is the relative permittivity/permeability of the filler material.

The electromagnetic energy is contained within the coax, so it is isolated from external effects. Because of its size, coax is not popular inside compact modules. However, it is popular in measurement benches as a way to connect gate and drain voltages while maintaining isolation from potential interfering microwave signals.

The design equations for coax lines are listed as follows. Wherever practical, approximations are included. Variables are shown in Figure 2.13 [2].

- Capacitance per unit length:

$$C = \frac{2\pi\varepsilon_o\varepsilon_r}{\ln\dfrac{D}{d}} F/m \approx \frac{55.556\varepsilon_r}{\ln\dfrac{D}{d}} pF/m \tag{2.25}$$

- Inductance per unit length:

$$L = \frac{\mu_o\mu_r}{2\pi}\ln\frac{D}{d} H/m \approx 200\ln\frac{D}{d} nH/m \tag{2.26}$$

- Characteristic line impedance:

Figure 2.13   Example coax cable.

$$Z_c = \sqrt{\frac{L}{C}} = \frac{1}{2\pi}\sqrt{\frac{\mu_o\mu_r}{\varepsilon_o\varepsilon_r}}\ln\frac{D}{d} \approx \frac{60}{\sqrt{\varepsilon_r}}\ln\frac{D}{d} \qquad (2.27)$$

- Propagation velocity (m/s):

$$v_p = \frac{c}{\sqrt{\varepsilon_r}} \qquad (2.28)$$

- Guided wavelength (m):

$$\lambda_g = \frac{c}{f\sqrt{\varepsilon_r}} \qquad (2.29)$$

- Propagation delay per unit length (ns/m):

$$\tau_{pd} = 3.33\sqrt{\varepsilon_r} \qquad (2.30)$$

- Attenuation due to dielectric loss per unit length in decibels:

$$\alpha_d = \frac{27.3\sqrt{\varepsilon_r}\,\tan\delta}{\lambda_o} \qquad (2.31)$$

- Cut-off frequency for utilization (Hz):

$$f_c = \frac{c}{\pi\left(\dfrac{D+d}{2}\right)\sqrt{\mu_r\varepsilon_r}} \qquad (2.32)$$

2.3.3 Microstrip

If the coax line is the most familiar structure, the microstrip line is arguably the most popular. Energy flows along a single conductor placed over a flat ground plane separated by a substrate. In Figure 2.14, the conductor has width w, length l, and thickness t. The substrate has thickness h, permittivity ε_r, permeability μ_r, and loss tangent (tan δ) [2].

Since only one substrate is needed, microstrip lines are easy and low-cost to manufacture and integrate with other components. Unlike with coax, there

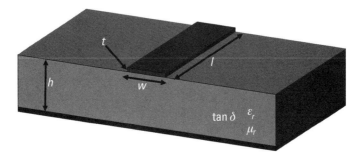

Figure 2.14 Example microstrip line.

is no cutoff frequency, but microstrip lines are principally used below 100 GHz. To make circuits smaller, they can be meandered or wrapped around other structures.

There are, however, drawbacks to microstrip lines, and the most critical is power handling. Excessive voltage levels will breakdown the substrate and arc to ground. Excessive current levels will heat the conductor line until it delaminates and breaks. Substrate properties change with temperature, and if not carefully screened, can have a 10% variability in manufacturing tolerance.

Some substrates are less sensitive to temperature change than others. Commercially available materials have values of ε_r that vary between approximately 3 and 500 ppm/°C (parts-per-million/°C). In applications where temperature will vary greatly, this must be a material-selection criteria.

Since the signal conductor is not surrounded on all sides by dielectric, the mode that propagates is not pure TEM. It is often referred as quasi-TEM in literature. This complicates the design equations, and approximations are available for a given range of width-to-height ratios. The effective dielectric constant can be approximated by the Schneider-Hammerstad equation [2]:

- For w/h ≤ 1:

$$\varepsilon_e = \frac{\varepsilon_r + 1}{2} + \frac{\varepsilon_r - 1}{2}\left(1 + \frac{12h}{w}\right)^{-0.5} + 0.02(\varepsilon_r - 1)(1 - w/h)^2$$

$$-0.217(\varepsilon_r - 1)\frac{t}{\sqrt{wh}} \tag{2.33a}$$

- For w/h > 1:

$$\varepsilon_e = \frac{\varepsilon_r + 1}{2} + \frac{\varepsilon_r - 1}{2}\left(1 + \frac{12h}{w}\right)^{-0.5} - 0.217(\varepsilon_r - 1)\frac{t}{\sqrt{wh}} \tag{2.33b}$$

Figure 2.15 shows the relationship between ε_r, ε_e, and Z (which of course determines w and h). As Z decreases, ε_e approaches ε_r. As ε_r increases, the effect on ε_e becomes more pronounced.

Capacitance per unit length [3]:

$$C_a = \begin{cases} \dfrac{2\pi\varepsilon_o}{\ln\left(\dfrac{8h}{w} + \dfrac{w}{4h}\right)} & \text{for } w/h \leq 1 \\[3ex] \varepsilon_o\left[\dfrac{w}{h} + 1.393 + 0.667\ln\left(\dfrac{w}{h} + 1.444\right)\right] & \text{for } w/h \leq 1 \end{cases} \tag{2.34}$$

Characteristic line impedance:

$$Z_c = \frac{1}{C_a}\sqrt{\frac{\mu_o\varepsilon_o}{\varepsilon_e}} \tag{2.35}$$

Characteristic line impedance (expanded and approximated) [4]:

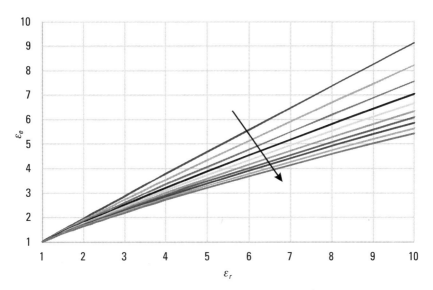

Figure 2.15 Relationship between ε_r, ε_e, and Z for a microstrip line (arrow denotes increasing Z from 10Ω to 100Ω)

$$Z_c = \begin{cases} 60\varepsilon_e^{-0.5} \ln\left(\dfrac{8h}{w} + \dfrac{0.25w}{h}\right) & \text{for } w/h \le 1 \\[3mm] \dfrac{120\pi\varepsilon_e^{-0.5}}{\dfrac{w}{h} + 1.393 + 0.667\ln\left(1.444 + \dfrac{w}{h}\right)} & \text{for } w/h > 1 \end{cases} \qquad (2.36)$$

Metal thickness is often specified to handle a desired current or power level (thicker metal handles more current and power). However, the fields propagating from thick metal layers are different than thin metal layers, so the width must be adjusted to compensate for the field change. The effects of metal thickness are included in the equation for ε_e, (2.33). To determine design rules, the metal thickness effect has been extracted and is represented in (2.37).

$$\Delta w = \begin{cases} 1.25\left(\dfrac{t}{\pi}\right)\left[1 + \ln\left(\dfrac{2h}{t}\right)\right] & \text{for } w/h > 1/(2\pi) \\[3mm] 1.25\left(\dfrac{t}{\pi}\right)\left[1 + \ln\left(\dfrac{4\pi w}{t}\right)\right] & \text{for } w/h \le 1/(2\pi) \end{cases} \qquad (2.37)$$

When the thickness is zero, Δw is also zero. For most substrate boards, $w/h > 1/(2\pi)$ which means Δw is only a function of the substrate height (h) and the metal thickness (t). Figure 2.16 plots Δw versus h at three different commercially available metal thicknesses.

Practical Note

From Figure 2.16 we can generate a rule of thumb: The change in width due to metal thickness is approximately twice the metal thickness.

An approximation for determining characteristic line impedance from a known geometry (ε_r, h, w, t) is available when w/h is between 0.1 and 3 [4]:

$$Z_c = \frac{87}{\sqrt{\varepsilon_r + 1.41}} \ln\left(\frac{5.98h}{0.8w + t}\right) \qquad (2.38)$$

An approximation for determining line width from a known substrate (ε_r, h, t) and desired characteristic line impedance is available when w/h is between 0.1 and 3 [4]:

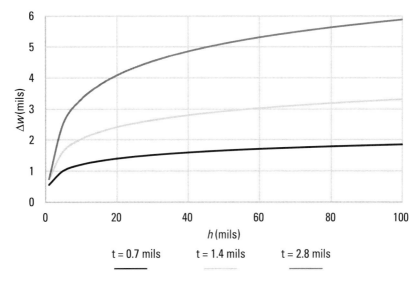

$t = 0.7$ mils $t = 1.4$ mils $t = 2.8$ mils

Figure 2.16 Relationship between of Δw, h, and t for a microstrip line (top line: 2.8 mils; middle line: 1.4 mils; bottom line: 0.7 mils).

$$w = \frac{7.48h}{e^{Z_o \frac{\sqrt{\varepsilon_r + 1.41}}{87}}} - 1.25t \tag{2.39}$$

Propagation delay per unit length (nanosecond per meter, ns/m):

$$\tau_{pd} = 3.33\sqrt{0.475\varepsilon_r + 0.67} \tag{2.40}$$

Propagation velocity:

$$v_p = \frac{c}{\sqrt{\varepsilon_e}} \tag{2.41}$$

Guided wavelength:

$$\lambda_g = \frac{c}{f\sqrt{\varepsilon_e}} \tag{2.42}$$

Losses during signal propagation come from the conductor (conductor loss), the substrate (dielectric loss), and radiation losses. At high frequency, conductor loss is much higher than dielectric and radiation [3].

The conductor loss can be calculated from:

$$\alpha_c = \frac{R_s + R_g}{2Z_c} \tag{2.43}$$

where R_s is the series resistance of the signal metal and R_g is the series resistance of the ground metal.

To calculate R_s and R_g, the resistance from the skin-effect (explained in Section 2.4.2) can be calculated from:

$$R_{skin} = \sqrt{\frac{2\pi f \mu}{\sigma}} \tag{2.44}$$

where f is frequency, μ is permeability, and σ is the metal conductivity (Siemens per meter, S/m).

R_g can be calculated from:

$$R_g = \frac{R_{skin}}{w} \frac{\frac{w}{h}}{\frac{w}{h} + 5.8 + 0.03\frac{w}{h}} \tag{2.45}$$

R_s can be calculated from:
For $w/h \leq 0.5$:

$$R_s = \frac{R_{skin}}{w}\left(\frac{1}{\pi} + \frac{1}{\pi^2}\ln\frac{4\pi w}{t}\right) \tag{2.46}$$

For $0.5 < w/h \leq 10$:

$$R_s = \frac{R_{skin}}{w}\left(\frac{1}{\pi} + \frac{1}{\pi^2}\ln\frac{4\pi w}{t}\right)\left(0.94 + 0.132\frac{w}{h} - 0.0062\left(\frac{w}{h}\right)^2\right) \tag{2.47}$$

A good approximation for the conductor loss in decibels per guide wavelength can also be calculated from the Hammerstad-Bekkadal equation:

$$\alpha_c \approx 0.072\frac{\sqrt{f}\lambda_g}{wZ_c} \tag{2.48}$$

The dielectric loss can be calculated from (units nepers/wavelength):

$$\alpha_d = \frac{\pi}{\lambda_o} \frac{\varepsilon_r}{\sqrt{\varepsilon_e}} \frac{\varepsilon_e - 1}{\varepsilon_r - 1} \tan \delta \tag{2.49}$$

Radiation loss comes from higher-order modes that propagate due to discontinuities that radiate energy. Discontinuities are discussed in Section 2.3.7. For high dielectric materials, surface wave propagation can also radiate energy and increase loss. Generally, if the substrate thickness is much less than $0.01 \times \lambda_o$, radiation loss is negligible.

Practical Note

Since propagation is quasi-TEM (as opposed to pure TEM), microstrip lines are often contained within metal or ferrite-loaded enclosures. To prevent disrupting the desired electric field, enclosures should be at least five times the substrate height (h) above the ground plane and two and a half times the conductor width (w) away from the conductor. This also applies to other circuit elements.

Microstrip structures have been modified to support differential operation (explained in Chapter 4). The structure, shown in Figure 2.17, differs from the standard microstrip line by the addition of a second conductor spaced apart by gap d.

An approximation for a differential microstrip impedance ($Z_{c,d}$) is available when w/h is between 0.1 and 3 [4]:

$$Z_{c,d} = \frac{174}{\sqrt{\varepsilon_r + 1.41}} \ln\left(\frac{5.98h}{0.8w + t}\right)\left(1 - 0.48e^{-0.96\frac{d}{h}}\right) \tag{2.50}$$

Microstrip structures can also be embedded within the substrate. This has the advantage of protecting the conducting material against surface damage (i.e., scratching or corrosion). The structure, shown in Figure 2.18, differs from the standard microstrip line by the addition of a layer of substrate above the conductor of thickness h_1.

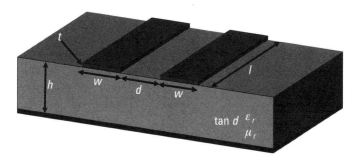

Figure 2.17 Example differential microstrip line.

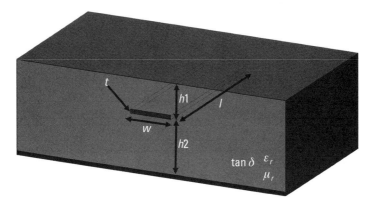

Figure 2.18 Example of embedded microstrip line.

An approximate for an embedded microstrip impedance ($Z_{c,e}$) is available when $(h_1+t+h_2)/h_2 > 1.2$ [4]:

$$Z_{c,e} = \frac{60}{\sqrt{\varepsilon_r \left(1 - e^{-1.55\frac{h1+t+h2}{h2}}\right)}} \ln\left(\frac{5.98h2}{0.8w + t}\right) \qquad (2.51)$$

$Z_{c,e}$ is plotted in Figure 2.19 with $t = 1.4$ mils, $h_2 = 10$ mils, and w is the equivalent for 50Ω. Figure 2.19 shows data sets for five commercially available substrates.

Practical Note

Interestingly, the relationship of ε_r to $Z_{c,e}$ is inverse parabolic. The ε_r that is least affected by the top-layer substrate is 6. The most affected materials have permittivity 2.2 and 10.

The propagation delay per unit length can be calculated from nanosecond per meter:

$$\tau_{pd} = 2.78\sqrt{\varepsilon_r \left(1 - \exp\left(-1.55\frac{h1 + t + h2}{h2}\right)\right)} \qquad (2.52)$$

Microstrip equations assume design parameters (i.e., Z_o and ε_e) are independent of frequency. In reality, as frequency increases, ε_e approaches ε_r. The effect of frequency on these parameters is called *dispersion*. The only way to truly determine the optimal line width for a given geometry is through full-wave

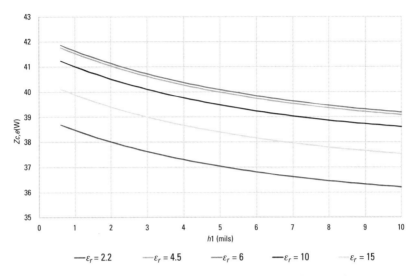

Figure 2.19 $Z_{c,e}$ versus $h1$ with $t = 1.4$ mils and $h2 = 10$ mils at various εr values.

analysis. Figure 2.20 shows the change in width that is required to compensate for dispersion maintaining 50Ω on a 20-mil-thick substrate with permittivity equal to 6. Dispersion is negligible below 9 GHz and then increases approximately linearly with a slope of 0.26 mil/GHz.

2.3.4 Stripline

Like a coax, the electromagnetic energy that flows through a stripline is contained within a grounded structure. Energy flows along a single conductor placed between two substrates with ground planes. In Figure 2.21, the conductor has width w, length l, and thickness t. The substrates have thickness h, permittivity ε_r, permeability μ_r, and loss tan δ [2].

Since dielectric surrounds the conductor, the stripline supports pure TEM propagation. Therefore, the effective dielectric constant is frequency-independent. Generally, stripline structures are used below 10 GHz. There is no dispersion, which makes stripline a popular option for directional couplers.

Calculating Z_c for a stripline is a multistep process. First, a parameter k is calculated from the width and height [3].

$$k = \sqrt{1 - \left(\tanh \pi \frac{w}{4h} \right)^2} \tag{2.53}$$

Second, Z_c can be calculated from k, w, h, and ε_r:

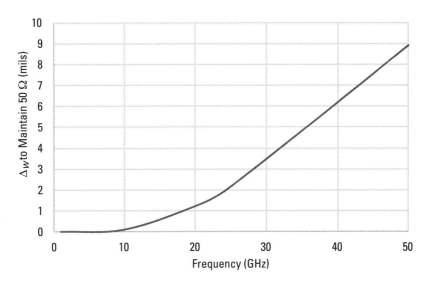

Figure 2.20 Change in width required to compensate for dispersion on a 20-mil-thick substrate with $\varepsilon_r = 6$

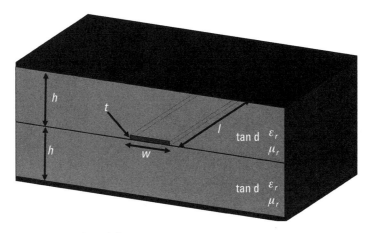

Figure 2.21 Example of a stripline.

$$Z_c = \frac{30\pi}{\sqrt{\varepsilon_r}} \cdot \begin{cases} \dfrac{1}{\pi} \ln\left(2 + \dfrac{1+\sqrt{k}}{1-\sqrt{k}}\right) & \text{for } 0.7 \leq k < 1 \\[4mm] \left[\dfrac{1}{\pi}\ln\left(2\dfrac{1+\sqrt{\tanh\pi\dfrac{w}{4h}}}{1-\sqrt{\tanh\pi\dfrac{w}{4h}}}\right)\right]^{-1} & \text{for } 0 < k \leq 0.7 \end{cases} \qquad (2.54)$$

Equation (2.54) assumes that the effects of metal thickness are negligible. For thick metal structures, the effects can be taken into consideration though another multistep process. First, a parameter x is determined from t and h [3]:

$$x = \frac{t}{2h}$$ (2.55)

Second, a parameter m is determined from x:

$$m = 2\left(1 + \frac{2}{3}\frac{x}{1-x}\right)^{-1}$$ (2.56)

Third, a parameter Ψ is determined from w, h, t, x, and m:

$$\Psi = \frac{w}{2h-t} + \frac{x}{\pi(1-x)}\left\{1 - \frac{1}{2}\ln\left[\left(\frac{x}{2-x}\right)^2 + \left(\frac{0.0796x}{w/2h+1.1x}\right)^m\right]\right\}$$ (2.57)

Finally, the stripline characteristic impedance can be calculated from Ψ and ε_r:

$$Z_c = \frac{30}{\sqrt{\varepsilon_r}}\ln\left\{1 + \frac{4}{\pi\Psi}\left[\frac{8}{\pi\Psi} + \sqrt{\left(\frac{8}{\pi\Psi}\right)^2 + 6.27}\right]\right\}$$ (2.58)

An approximate for a stripline characteristic impedance (Z_c) is available when w/h is 0.1-2 and $t/h < 0.25$:

$$Z_c = \frac{60}{\sqrt{\varepsilon_r}}\ln\left(\frac{1.9(2h+t)}{0.8w+t}\right)$$ (2.59)

The propagation delay per unit length nanosecond per meter, ns/m can be calculated from:

$$\tau_{pd} = 3.33\sqrt{\varepsilon_r}$$ (2.60)

The dielectric loss (nepers/wavelength) can be calculated from:

$$\alpha_d = \frac{\pi}{\lambda_o}\sqrt{\varepsilon_r}\tan\delta$$ (2.61)

The conductor loss can be calculated from:

$$\alpha_c = \alpha_s + \alpha_g \tag{2.62}$$

where α_s is the conductor loss of the signal metal, and α_g is the conductor loss of the ground metal.

α_s and α_g can be calculated by [3]:

$$\alpha_s = \frac{\pi R_{skin}}{16 Z_c h K'^2 \tanh\left(\pi \frac{w}{4h}\right)} \ln \frac{16h \tanh\left(\pi \frac{w}{4h}\right)}{\frac{\pi}{\cosh\left(\frac{\pi w}{4h}\right)} e^{\frac{-\pi}{2}\sqrt{\frac{4wt}{\pi}}}} \tag{2.63}$$

$$\alpha_g = \frac{\pi^2 R_{skin} w}{64 Z_c h^2 K'^2 \tanh\left(\pi \frac{w}{4h}\right)} \tag{2.64}$$

where

$$K' = \int_0^{\pi/2} \frac{dx}{\sqrt{1 - \tanh^2\left(\pi \frac{w}{4h}\right) \sin^2 x}} \tag{2.65}$$

α_s and α_g can be approximated by [3]:

$$\alpha_s = \frac{R_{skin}\sqrt{\varepsilon_r}}{2 Z_c \cdot h} \frac{\ln\left(\frac{8h}{\pi e^{\frac{-\pi}{2}\sqrt{\frac{4wt}{\pi}}}}\right) + \frac{\pi w}{4h}}{\ln 2 + \frac{\pi w}{4h}} \qquad \text{for } w \geq 2h \tag{2.66}$$

$$\alpha_g = \frac{\pi w R_{skin}\sqrt{\varepsilon_r}}{8 Z_c \cdot h^2 \left(\ln 2 + \frac{\pi w}{4h}\right)} \qquad \text{for } w \geq 2h \tag{2.67}$$

$$\alpha_s = \frac{2R_{skin}\sqrt{\varepsilon_r}\ln\left(\dfrac{4w}{\exp\left(\dfrac{-\pi}{2}\sqrt{\dfrac{4wt}{\pi}}\right)}\right)}{\pi Z_c \cdot w\left(\ln\dfrac{16h}{\pi w}\right)} \qquad \text{for } w \leq 0.4h \tag{2.68}$$

$$\alpha_g = \frac{R_{skin}\sqrt{\varepsilon_r}}{2Z_c \cdot h\left(\ln\dfrac{16h}{\pi w}\right)} \qquad \text{for } w \leq 0.4h \tag{2.69}$$

Cutoff frequency for utilization (units are gigahertz if w and h are in centimeters):

$$f_c = \frac{15}{2h\sqrt{\varepsilon_r}}\frac{1}{\dfrac{w}{2h}+\dfrac{\pi}{4}} \tag{2.70}$$

Stripline structures have been modified to support differential operation. The structure is shown in Figure 2.22 and differs from the standard stripline line by the addition of a second conductor spaced apart by gap d.

An approximation for a differential stripline impedance ($Z_{c,d}$) is available when w/h is 0.1-2 and $t/h < 0.25$ [4]:

$$Z_{c,d} = \frac{120}{\sqrt{\varepsilon_r}}\ln\left(\frac{1.9(2h+t)}{0.8w+t}\right)\left(1-0.347e^{-2.9\frac{d}{2h+t}}\right) \tag{2.71}$$

Equation (2.71) assume the substrate thickness above and below the conductor are the same (h). A variation where this is not the case is shown in Figure 2.23.

An approximate for an asymmetric stripline impedance can be calculated when w/h is 0.1-2 and $t/h < 0.25$ [4]:

$$Z_{c,a} = \frac{80}{\sqrt{\varepsilon_r}}\ln\left(\frac{1.9(2h1+t)}{0.8w+t}\right)\left(1-\frac{h1}{4h2}\right) \tag{2.72}$$

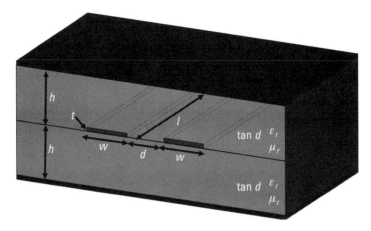

Figure 2.22   Example differential stripline.

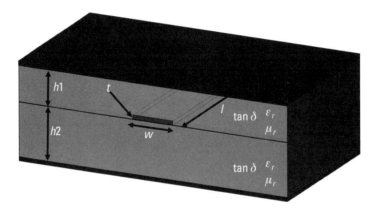

Figure 2.23   Example asymmetric stripline.

The propagation delay per unit length (ns/m) can be calculated from:

$$\tau_{pd} = 3.333\sqrt{\varepsilon_r} \tag{2.73}$$

2.3.5   Coplanar Waveguide (CPW)

A CPW requires only one substrate like a microstrip line, so it has the same cost and manufacturing advantages. Energy flows along a single conductor placed between two ground planes on the same side of the substrate. In Figure 2.24, the conductor has width w, ground width D, gap g, and thickness t. The substrate has thickness h, permittivity ε_r, permeability μ_r, and loss tan δ [2].

Figure 2.24   Example of a coplanar waveguide.

Since ground is on the same side as the conductor, there is no need for vias to ground. This makes incorporating shunt circuits much easier. Additionally, CPW lines have low dispersion and are commonly used to 60 GHz.

Practical Note

There is flexibility in the required ground width D. A good rule of thumb is: $\lambda/2 > 2D + 2g + w > 10(w + 2g)$.

As with microstrip lines, waves propagate within two materials in a CPW so an effective dielectric constant must be determined:

$$\varepsilon_e = \frac{\varepsilon_r + 1}{2} \tag{2.74}$$

Calculating the characteristic impedance of a CPW requires multiple steps. First, a parameter k must be calculated based on w and g:

$$k = \frac{w}{w + 2g} \tag{2.75}$$

Second, the characteristic impedance can be calculated from k and ε_e:

$$Z_c = \frac{30\pi^2}{\sqrt{\varepsilon_e}} \frac{1}{\ln\left(2 + \dfrac{1 + \sqrt{k}}{1 - \sqrt{k}}\right)} \tag{2.76}$$

The dielectric loss can be calculated using the microstrip equation (units nepers/wavelength):

$$\alpha_d = \frac{\pi}{\lambda_o} \frac{\varepsilon_r}{\sqrt{\varepsilon_e}} \frac{\varepsilon_e - 1}{e_r - 1} \tan \delta \qquad (2.77)$$

The conductor loss can be calculated from:

$$\alpha_c = \frac{R_s + R_g}{2Z_c} \qquad (2.78)$$

where R_s is the series resistance of the signal metal and R_g is the series resistance of the ground metal.

R_s and R_g can be calculated from [3]:

$$R_s = \frac{R_{skin}}{4w\left(1 - k^2\right)K^2(k)} \left(\pi + \ln\frac{4\pi w}{t} - k \ln\frac{1+k}{1-k} \right) \qquad (2.79)$$

$$R_g = \frac{k \cdot R_{skin}}{4w\left(1 - k^2\right)K^2(k)} \left(\pi + \ln\frac{4\pi\left(w + 2g\right)}{t} - \frac{1}{k} \ln\frac{1+k}{1-k} \right) \qquad (2.80)$$

Practical Note

To minimize dispersion effects and suppress unwanted mode propagation, the ground lines should be connected by wire bonds throughout the structure and at every discontinuity (i.e., bends). Figure 2.25 shows an example of resolved discontinuity.

2.3.6 Waveguide

Unlike the structures we've discussed so far, waveguides are not fabricated on a substrate. Instead, they are machined from metal or another rigid material (i.e., silicon) and then metalized. Figure 2.26 shows a rectangular waveguide with width a, height b, length l, permittivity ε_r, and permeability μ_r [2].

Waveguides have the advantages of very low loss and a high power capability, and they support high-Q (narrow bandwidth relative to the center frequency) passive circuits. They are also physically large, expensive, and difficult to integrate. For these reasons, waveguides are not the focus of this book. Nevertheless, this section presents the important waveguide equations.

Figure 2.25 Bond wires added to resolve CPW discontinuity.

Figure 2.26 Example of a rectangular waveguide.

Since there is no center conductor, TEM waves cannot propagate. Instead TE_{nm} and TM_{nm} modes are present. For a waveguide, n can be 0, 1, 2, … and m can be 0, 1, 2, …, but $n \neq m = 0$. TE10 is generally the preferred mode. Figure 2.27 shows what the TE and TM modes look like within the waveguide. That pattern repeats throughout the length of the structure, but in alternating direction each cycle.

The propagation constant γ_{nm} can be calculated by:

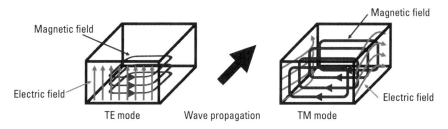

Figure 2.27 TE and TM modes within a waveguide.

$$\gamma_{nm=}\sqrt{\left(\frac{n\pi}{a}\right)^2 + \left(\frac{m\pi}{b}\right)^2 - \left(2\pi f\right)^2 \mu_o \mu_r \varepsilon_o \varepsilon_r} \qquad (2.81)$$

The minimum cutoff frequency (waves only propagate if $f > f_{c,nm}$) is calculated from:

$$f_{c,nm} = \frac{c}{2\pi\sqrt{\mu_r \varepsilon_r}}\sqrt{\left(\frac{n\pi}{a}\right)^2 + \left(\frac{m\pi}{b}\right)^2} \qquad (2.82)$$

The guide wavelength is

$$\lambda_{g,nm=}\frac{1}{\sqrt{\mu_r \varepsilon_r}}\frac{\lambda_o}{\sqrt{1-\left(\frac{f_{c,nm}}{f}\right)^2}} \qquad (2.83)$$

The wave impedance for TE$_{nm}$ mode is

$$Z_{TE,nm} = \lambda_{g,nm} f \mu_o \mu_r \qquad (2.84)$$

The wave impedance for TM$_{nm}$ mode is

$$Z_{TM,nm} = \frac{1}{\lambda_{g,nm} f \varepsilon_o \varepsilon_r} \qquad (2.85)$$

2.3.7 Discontinuities

The equations included in this chapter apply to straight, continuous, and uninterrupted transmission lines. Whenever a deviation from this structure is introduced, a potential discontinuity is introduced. This can be in the form of bends, gaps, width changes, thickness changes, and material changes, among others. Discontinuities can be purposely introduced for filtering or impedance matching. Often, they are added to reduce size or material cost. Depending on the structure, they can introduce an unwanted inductance or capacitance that can limit bandwidth and degrade performance.

Figure 2.28 shows the parasitic effects of four common discontinuities with their respective circuit model. Chapter 6 presents methods for mitigating these discontinuities.

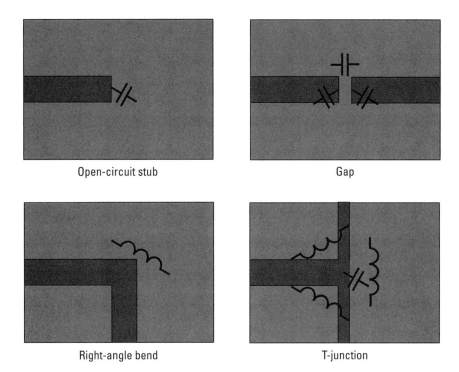

Open-circuit stub　　　　　　　　Gap

Right-angle bend　　　　　　　　T-junction

Figure 2.28　Parasitic effects of four common discontinuities.

2.4　Material Selection

It probably goes without saying that material selection is critical in every aspect and discipline of engineering. No one would ever make a bridge out of paper mâché or an airplane out of brick. Materials must be chosen based on the following factors:

- Physical properties (i.e., melting point and reflectivity);
- Electrical properties (i.e., conductivity, dielectric constant, and loss tangent);
- Mechanical properties (i.e., Poisson's ratio and fatigue resistance);
- Thermal properties (i.e., thermal expansion, specific heat, and conductivity);
- Chemical properties (i.e., corrosion, radiation, and oxidation resistance);
- Manufacturing properties (i.e., machinability and available tolerances);
- Cost (i.e., whether it meets the financial requirements);
- Availability (i.e., whether it can be procured in the time allowed);

• Environmental impact (i.e., whether it contains lead).

Guidelines for material selection will be discussed throughout the book. It is a design parameter that is especially critical for radars due to these inherent characteristics of radar:

- Concentration of heat in small areas;
- Fluctuation of electrical and environmental conditions;
- Requirements for cost reduction;
- Low tolerance for failure;
- Long expected lifetime;
- Severe operating conditions.

This section provides an overview of materials commonly used in component design. The appendix lists tables of material properties.

2.4.1 Semiconductors

Semiconductors provide the backbone for monolithic microwave integrated circuits (MMICs), which are the RF equivalent of the integrated circuits found in everyday digital chips. Common semiconductors used in microwave radar components include Gallium Arsenide (GaAs) and Gallium Nitride (GaN). Semiconductors tend to have the following material characteristics [5]:

- They are poor conductors of electricity;
- They have a high temperature stability;
- Their purity can be implemented to within a few impurity atoms for every billion host atoms (parts-per-billion, ppb);
- They are chemically resistant.

2.4.2 Metals

Metals are both electrically and thermally conductive. They are used as the conductive material for signal and ground lines. Depending on the application, metal lines can be deposited by sputtering, evaporating, electroplating, or rolling. They are also used as the carrier or base plate metal to spread heat and add mechanical stability. Common metals used in microwave radar components are gold (Au), copper (Cu), and aluminum (Al). Metals tend to offer the following material properties [5]:

- They are excellent conductors of both electricity and heat;
- They are relatively strong;
- They are highly dense;
- They are malleable;
- They are resistant to cracking;
- They are resistant to breaking when subjected to high-impact forces.

Microwave signals do not propagate equally throughout a volume of conductive material. The amplitude of the field as it decays within a conductor can be calculated by:

$$A_d = e^{-d/\delta_s} \tag{2.86}$$

where A_d is the magnitude of field strength present (maximum is 1 or 100% field remaining), d is the depth into the material from the surface (m), and δ_s is the skin depth (m).

The skin depth is the depth within a conductive surface where most of the current flows, and it is calculated by:

$$\delta_s = \sqrt{\frac{1}{\pi f \mu \sigma}} \tag{2.87}$$

where f is frequency, μ is permeability, and σ is conductivity loss. More than one-third of the propagating signal is absorbed at depths beyond the skin depth so it is important to choose a metal and metal thickness sufficient to support the radar signals needed.

2.4.3 Ceramics

Ceramics are available with a wide array of properties. For example, thermal conductivity ranges from nearly perfectly insulating (i.e., space shuttle tiles during re-entry) to more conductive than metal (i.e., carbon-based nanocomposites). Ceramics tend to offer the following material properties [5]:

- They are generally composed of both metallic and nonmetallic elements;
- They are poor conductors of electricity;
- Their thermal conductivity can range from very poor to superior to metal;
- They are stronger than metal, but brittle under impact force;

- They are high-temperature-stable;
- They are chemically resistant.

2.4.4 Polymers

Polymers are generally used as low-cost packaging and sealing materials. Many are specifically designed for high-frequency or military applications. Although there are many exceptions, polymers tend to offer the following material properties [5]:

- They soften at high temperature;
- They are inexpensive to make;
- They are low-density;
- They are malleable;
- They are chemically resistant.

2.4.5 New and Emerging Technologies

New materials have brought new life to old (perhaps stale) technologies. Companies are taking existing designs and porting them to new materials with improvements like lower loss, higher permittivity, lower cost, better uniformity, and improved robustness. New designs can leverage the advancements that have been made and that are being made every day. Some of the most exciting new and emerging materials are described as follows.

Electrically Conductive Polymers

Polymers provide low-cost and low-loss packaging options. Unfortunately, since they are nonmetallic, they do not provide shielding (discussed in Chapter 8). In highly integrated assemblies, this can cause crosstalk among components. Electrically conductive polymers are being introduced that provide the moldability and cost savings of polymers with the shielding properties of metal. Several polymers, including polyacetylene, polyparaphenylene, polypyrrole, polyaniline, and polythiophene, can be made conductive. Alternatively, other polymers can be made conductive by filling with conductive materials (i.e., silicone rubber or silver particles).

Liquid Crystal Polymer (LCP)

LCP has a number of unique properties. First, it operates with very low loss up to 110 GHz making it applicable for millimeter-wave applications. Second, it is hydrophobic so moisture is not absorbed into the material. This allows it to maintain electrical properties even in humid environments. Third, it is equally qualified to be both a substrate and a packaging material. Commercial fabricators can now assemble structures with more than 10 layers with precision. Fourth, it is a flexible material that has been used to make conformal electronics.

High-Permittivity Substrates

Circuit size is inversely proportional to the permittivity. Circuits are inherently smaller on higher-permittivity substrates than lower-permittivity substrates. As the need for smaller components increases, the use of these high-permittivity substrates becomes more prevalent. Materials doped with hafnium and beryllium are gaining in popularity and can have permittivity of 30–100. The drawback to increased permittivity is usually higher loss.

Robust Substrates

Companies that produce microwave substrates are making materials geared toward military applications. Some offer temperature stability (very little change in performance versus temperature). Others, like LCP, operate well in harsh environments. The best manufactures make these changes without sacrificing loss (tan δ) or cost.

Diamond

When thermal management is critical and cost is secondary, there is no better heat spreader on the market than diamond. Sheets of diamond are commercially available with thermal conductivity up to a staggering 1,800 W/m·K. This is nearly five times more conductive than pure copper. Diamond is electrically an insulator so before using it as a thermal spreader, all sides must be metalized to preserve the ground plane.

Exercises

1. A 50-Ω transmission line is connected to a 35-Ω load.
   - What is the reflection coefficient?
   - What is the input return loss?
   - What is the VSWR?
   - If a 5-W signal is incident to the transmission line, how much power will be reflected from the load?
2. Design a matching network from 30+j50Ω to 50Ω using as few elements as possible.

3. A Teflon-filled coax has an inner diameter of 3.6 mm and an outer diameter of 5.08 mm. Using the material properties listed in the appendix:
   - What is the capacitance per unit length?
   - What is the inductance per unit length?
   - What is the characteristic line impedance?
   - What is the cutoff frequency?

4. To meet a desired VSWR specification, the impedance of a microstrip line on 25-mil thick LCP must be $50 \pm 20\Omega$. How thick of an LCP layer can be placed on top of the microstrip line and still meet this requirement?

5. What range of widths is suitable for the ground pads of a 12-GHz coplanar waveguide on 4-mil-thick gallium arsenide?

6. How much metal is required to achieve three skin depths thickness for aluminum, gold, and silver?

7. How does the minimum TE10 cutoff frequency of a rectangular waveguide change if:
   - The width doubles?
   - The height doubles?
   - The length doubles?

References

[1]   Pozar, D., *Microwave Engineering*, New York, NY: Wiley, 1997.

[2]   Chang, K., I. Bahl, and V. Nair, *RF and Microwave Circuit and Component Design for Wireless Systems*, New York, NY: Wiley, 2002.

[3]   Collin, R., *Foundations for Microwave Engineering*, New York, NY: IEEE Press, 2001.

[4]   Mantaro Product Development Services, Inc., "Impedance Calculators," Internet, http://www.mantaro.com/resources/impedance_calculator.htm.

[5]   Schaffer, J., et al., *The Science and Design of Engineering Materials*, New York, NY: WCB McGraw-Hill, 1999.

Selected Bibliography

Maas, S., *Practical Microwave Circuits*, Norwood, MA: Artech House, 2014.

Komarov, V., *Handbook of Dielectric and Thermal Properties of Materials at Microwave Frequencies*, Norwood, MA: Artech House, 2012.

Garg, R., I. Bahl, and M. Bozzi, *Microstrip Lines and Slotlines,* Norwood, MA: Artech House, 2013.

Matthaei, G., L. Young, and E. Jones, *Microwave Filters, Impedance-Matching Networks, and Coupling Structures,* Norwood, MA: Artech House, 1980.

Joines, W., W. Palmer, and J. Bernhard, *Microwave Transmission Line Circuits,* Norwood, MA: Artech House, 2013.

3

Component Modeling

The behavior of all microwave components, including those listed in Table 1.2, can be expressed mathematically. Behavior can be categorized by the frequency response (how it behaves over frequency), transient response (how it behaves over time), small-signal response (how it behaves at low input power level), and large-signal response (how it behaves at high input power level).

For complicated components or components with many ports, working with a system of equations to model the behavior can be cumbersome. To simplify the design process, equivalent circuit models are derived using basic electrical elements, including capacitors, inductors, resistors, transmission lines (discussed in Chapter 2), voltage sources, and current sources. Each of those basic elements can be described by a much more manageable set of equations.

This chapter presents equivalent circuit models that describe the electrical behavior of components at microwave frequencies, including unwanted effects caused by the physical structure (i.e., packaging effects). Those models are then combined to generate a circuit model for a transistor—arguably one of the most complicated components used by a microwave designer. The model serves as an example to explain some of the physical traits of a transistor, which will be leveraged in Chapters 4 and 5 (amplifier design). The process for deriving custom models is also explained.

3.1 Passive Modeling

Lumped or surface-mount elements are attractive due to their small size, broad bandwidth, and commercial availability compared with distributed elements. At low frequencies, component size is very small compared to the wavelength. Therefore, they can be used with minimal concern. However, at higher frequencies (UHF and above), this is not the case. Lumped elements exhibit spurious resonances, fringing fields (radiation leakage), loss, and other parasitic (undesired) effects (discussed in Chapter 6) [1].

3.1.1 Capacitor

The equivalent circuit model of a lumped-element capacitor is shown in Figure 3.1 where C_{nom} is the nominal capacitance, L_s is the series parasitic inductance, ESR is the equivalent series resistance, and C_p is the parallel parasitic capacitance.

Fortunately, data sheets often include one or more of these parameters to aid designers. If not, S-parameters are usually available so that values can be tuned to match the frequency response. Figure 3.2 shows the effect L_s has on the capacitor frequency response. The $L_s = 0.000$ curve has no parasitic effect. As L_s increases, the high-frequency response degrades quickly.

Some components are marketed as having "low ESR." They are designed for high-frequency applications. Having a low ESR will reduce the loss, especially at high frequency.

Sometimes data sheets quantify the loss of capacitor by its quality factor (Q), which is defined as:

$$Q = \frac{1}{2\pi f C_{nom} \cdot ESR} \tag{3.1}$$

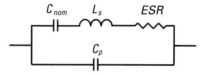

Figure 3.1 Equivalent circuit model of a real capacitor.

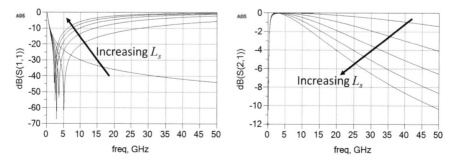

Figure 3.2 Effect of parasitic L_s (0–1 nH) on 5-pF capacitor frequency response

where f is the frequency (hertz), C_{nom} is the nominal capacitance (farads), and ESR is the series resistance (Ω). This equation can be used to calculate ESR.

Data sheets may also specify the series-resonant frequency (SRF) of a capacitor. This is the frequency where C_{nom} and L_s are equal, but opposite in magnitude. At that frequency, the capacitor behaves like a parallel RC circuit with ESR and C_p. To calculate L_s from SRF, the following equation can be used:

$$L_s(\text{nH}) = \frac{1000}{4\pi^2 C(\text{pF}) \cdot \text{SRF}(\text{GHz})^2} \tag{3.2}$$

C_p accounts for any deviation from the frequency response that cannot be predicted. It is determined by trial and error until the measured frequency response matched the equivalent circuit model.

3.1.2 Inductor

The equivalent circuit model of a lumped-element inductor is shown in Figure 3.3 where L_{nom} is the nominal inductance, ESR is the parallel resistance, and C_p is the parallel parasitic capacitance.

Sometimes data sheets quantify the loss of inductor by its quality factor (Q), which is defined as:

$$Q = \frac{2\pi fL}{R} \tag{3.3}$$

where f is the frequency (Hz), L is the inductance (henrys), and R is the series resistance (Ω). This equation can be used to calculate R. R can also be measured directly with an ohmmeter. If the inductor is comprised of a coil of wire, R can be approximated from the resistivity of the metal:

$$R = \frac{R_{skin}l}{\pi d} \tag{3.4}$$

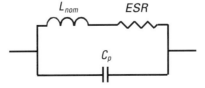

Figure 3.3 Equivalent circuit model of a real inductor.

where R_{skin} is the skin-effect resistance (2.44), l is the wire length (approximately equal to the number of turns multiplied by the coil circumference plus the lead length), and d is the wire diameter.

Data sheets may also specify the *SRF* of an inductor. To calculate C_p from *SRF*, the following equation may be used:

$$C_p(\text{pF}) = \frac{1000}{4\pi^2 L(\text{nH}) \cdot \text{SRF}(\text{GHz})^2} \tag{3.5}$$

C_p accounts for any deviation from the frequency response that cannot be predicted. It is determined by trial and error until the measured frequency response matched the equivalent circuit model.

Wire bonds used to connect components together are another source of inductance in a circuit. There are many models that approximate the behavior of a wire. A simple, but popular model that determines the inductance of a straight wire based on physical parameters is [2]:

$$L(nH) = 2x10^{-4} l \left(\ln\frac{4l}{d} + 0.5\frac{d}{l} - 0.75 \right) \tag{3.6}$$

where l is the wire length (μm) and d is the wire diameter (μm).

When the wire is used on the surface of a substrate, the proximity to the ground plane (discussed in Chapter 2) can have an effect. This is particularly true for long wires or thin substrates (less than 5 mils). Equation (3.7) shows the added complexity when ground effects are considered [2].

$$L(nH) = 2x10^{-4} l \left[\ln\frac{4h}{d} + \ln\left(\frac{l + \sqrt{l^2 + \frac{d^2}{4}}}{l + \sqrt{l^2 + 4h^2}}\right) + \sqrt{1 + \frac{4h^2}{l^2}} - \sqrt{1 + \frac{d^2}{4l^2}} - 2\frac{h}{l} + \frac{d}{2l} \right] \tag{3.7}$$

where h is the distance above ground plane (μm), l is the wire length (μm), and d is the wire diameter (μm).

These equations do not take into account the height or shape of the bond wire arch. Design software packages include sophisticated bond wire models that take into consideration the substrate properties and geometry of the bond wire.

> **Practical Note**
>
> For a 1-mil wire, the equivalent inductance is largely related to substrate thickness and the bond wire height (the apex of the arch). Figure 3.4 shows a plot demonstrating a good rule of thumb relationship.

The inductance of a wire can be reduced by replacing it with a wider structure called a *ribbon*. The inductance of a ribbon can be calculated from [2]:

$$L(\text{nH}) = 0.2l\left(\ln\left(\frac{2l}{w+t} \right) + \frac{0.223(w+t)}{l} + 0.5 \right) \tag{3.8}$$

where l is the wire length (mm), w is the wire width (mm), and t is the wire thickness (mm).

For a single-layer air-wound coil inductor where the length of the coil is greater than one-half the diameter, the inductance can be calculated from [2]:

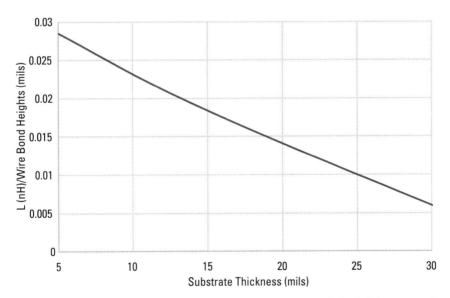

Figure 3.4 Rule of thumb relationship between inductance per bond wire height versus substrate thickness.

$$L(\text{nH}) = \frac{d^2 N^2}{1.016l + 0.4572d} \tag{3.9}$$

where d is the distance between wire centers in millimeters (inner diameter + wire diameter), N is the number of turns, and l is the length (millimeters).

Wires are generally specified in terms of gauge [American wire gauge (AWG)] instead of diameter. To convert from AWG to diameter:

$$D(\text{mm}) = 0.127 \cdot 92^{\frac{36 \cdot AWG}{39}} \tag{3.10}$$

3.1.3 Resistor

The equivalent circuit model of a lumped-element resistor is shown in Figure 3.5 where R_{nom} is the nominal resistance, L_s is the series parasitic inductance, and C_p is the parallel parasitic capacitance.

As is the case with capacitors and inductors, resistors have a frequency range where the parasitics overcome the resistance. L_s and C_p can be determined if SRF information is provided from the vendor, or through trial and error until the measured frequency response matches the model.

3.1.4 Resonators

Resistors, inductors, and/or capacitors can be combined in series or parallel to make circuits with very sharp frequency responses. A circuit might pass all frequencies except for a very narrow band or might reject all frequencies except for a very narrow band. Resonators are widely used in filters, so they will be discussed extensively in Chapter 6. Since they are also used in amplifiers and matching networks, their important design equations are included in this section.

The frequency of both series and parallel resistor, inductor, and capacitor (RLC) circuits is determined by:

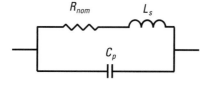

Figure 3.5 Equivalent circuit model of a real resistor.

$$f = \frac{1}{2\pi\sqrt{LC}} \tag{3.11}$$

Other circuit parameters are specific to the RLC orientation.

Series RLC Circuit

The bandwidth of the series circuit is defined by the resistor and inductor (solving for the difference in the half-power frequencies shows that the capacitor terms cancel):

$$BW = \frac{R}{L} \tag{3.12}$$

where BW is the bandwidth (Hz), R is the resistance (Ω), and L is the inductance (H, which is equivalent to $\Omega\cdot s$).

The quality factor (Q) quantifies how quickly the magnitude of the impedance changes with respect to frequency, and is calculated by:

$$Q = \frac{1}{R}\sqrt{\frac{L}{C}} = \frac{2\pi fL}{R} = \frac{1}{2\pi fCR} \tag{3.13}$$

The equivalent admittance can be calculated from:

$$Y = \frac{1}{R + j2\pi fL + \dfrac{1}{j2\pi fC}} \tag{3.14}$$

Therefore, the equivalent impedance can be calculated from:

$$Z = R + j2\pi fL - \frac{j}{2\pi fC} \tag{3.15}$$

Parallel RLC Circuit

The bandwidth of the parallel circuit is defined by the resistor and capacitor (solving for the difference in the half-power frequencies shows that the inductor terms cancel):

$$BW = \frac{1}{RC} \qquad (3.16)$$

where BW is the bandwidth (Hz), R is the resistance (Ω), and C is the capacitance (Farads, which is equivalent to s/Ω).

The quality factor is calculated by:

$$Q = R\sqrt{\frac{C}{L}} = \frac{R}{2\pi fL} = 2\pi fCR \qquad (3.17)$$

The equivalent impedance can be calculated from:

$$Z = \frac{1}{\dfrac{1}{R} + j2\pi fC + \dfrac{1}{j2\pi fL}} \qquad (3.18)$$

3.2   Footprint Modeling

As soon as a component is attached to a transmission line by way of soldering or epoxying to a pad of metal, the effects of the attachment must be considered. This is generally modeled as a set of pads (called the *footprint*) spaced the same distance apart as the component ports. Table 3.1 lists typical pad sizes along with their respective approximate capacitance (footprints vary by manufacturer). The frequency where the pad exhibits behavior other than a pure

Table 3.1
Typical Pad Sizes and Their Equivalent Capacitance and
Maximum Low-Parasitic Frequency

Package Size	Pad Width (mils)	Pad Length (mils)	Gap Length (mils)	Equivalent Capacitance (fF)*	Max Low-Parasitic Frequency (GHz)*
0201	5.9	11.8	11.8	7	55
0402	9.8	19.7	19.7	5	40
0603	11.8	31.5	39.4	1	40
0805	15.7	49.2	47.2	0.7	18
1206	19.7	63.0	86.6	0.022	10

* 20-mil-thick, $\varepsilon_r = 6$ substrate

capacitance (i.e., a parasitic inductance has been introduced) is the maximum low-parasitic frequency.

Chapter 6 discusses rules for laying out circuits to prevent coupling or set a desired level of coupling.

3.3   Transistor Modeling

The heart of an active circuit is the transistor. It is the component that uses DC power to amplify an RF signal. Chapter 2 discussed how a poor match can add loss. Transistors have an additional layer of complexity. The impedance presented at the input and output has a profound effect on the behavior of the transistor (this is discussed in Chapter 4). Additionally, its performance is determined by the DC bias applied and level of input RF power. To be effective, a transistor model must accurately represent this multidimensional trade space. This section reviews pertinent information about semiconductors and transistor operation.

3.3.1   Semiconductor Background

Materials of interest for transistors (and other electronics) can be categorized as either conductors, insulators, or semiconductors. Conductors are generally metals, and have resistivity less than 10^{-3} Ω·cm. Silicon dioxide is an example of an insulator used commonly in silicon transistors. Insulators can have resistivity of more than 10^5 Ω·cm. Semiconductors, as the name implies, are all materials between conductors and insulators. The resistivity of a homogeneous material (one type of material is used throughout the object) is determined by:

$$\rho = \frac{m}{e^2 n \tau} \tag{3.19}$$

where ρ is the resistivity (Ω·m), m is the electron mass (9.11×10^{-31} kg), e is the fundamental charge (1.602×10^{-19} C), n is the number of charge carriers per unit volume, and τ is the mean time between collisions of the charge carriers. The resistivity of pure copper is 1.7×10^{-8} Ω·m. The resistivity of pure gold is 2.4×10^{-8} Ω·m. Gold is approximately 40% more resistive than copper, which is why copper traces have gold plating (to prevent oxidation) is the preferred metal configuration for microwave components. The appendix presents a table of resistivity.

Nearly all RF transistors used in radars today are compound semiconductors, meaning they are formed by two or more elements. Generally, the most common transistors come from elements in the third and fifth columns of the periodic table so they are known as *III-V* (pronounced "three five") *transistors*. The most popular semiconductors for radar are gallium arsenide (GaAs), gallium nitride (GaN), and indium phosphide (InP). The advantages of these materials will be discussed in Part II.

A semiconductor is often classified by its *bandgap energy* (or just *bandgap*). It is known from quantum mechanics that electrons surround the atom at various discrete bands or energy states. Energy states where the electrons are bound to a particular atom are called *valence bands*. Energy states outside the valance band where electrons can flow freely from atom to atom within a lattice are called *conduction bands*. The energy differential between the lowest conduction band and the highest valence band is the bandgap. A perfect metal would have zero bandgap, and a perfect insulator would have infinite bandgap. In general, semiconductors have bandgaps less than 2.5 eV and insulators have bandgaps greater than 2.5 eV [3].

Practical Note

Bandgap values vary from foundry to foundry, so always ask before doing any kind of analysis that requires this information. The bandgap of GaN, for example, can vary from 1.5 to 3 eV for commercial processes. Chapter 7 discusses how this difference can have a profound effect on reliability and stability. For example, the difference can change the necessary burn-in time (operational time required before the performance of an active component remains constant) by more than 20 times (see Chapter 7).

Charge-carrier mobility, conductivity, and the number of charge carriers vary with temperature. For these reasons, the effects of temperature play an important role in transistor behavior and should be included in any model.

3.3.2 Basic Transistor Theory Review

At the simplest level, transistors generate an electric field from a DC bias and when an RF signal passes through that electric field, it is amplified. There are many different types of transistors, each with strengths and weaknesses. Several transistors with their electrical symbol are listed as follows:

- Bipolar junction transistors (BJTs);
- Heterojunction bipolar transistors (HBTs);
- Metal-oxide semiconductor field effect transistors (MOSFETs);
- Laterally diffused metal oxide semiconductors (LDMOS);
- Metal semiconductor field effect transistors (MESFETs);
- High electron mobility transistors (HEMTs).

BJTs utilize electrons and holes (absence of an electron) as charge carriers. Regions concentrated with electrons and holes are created by *doping* a semiconductor with impurities. When a BJT is biased, the charge carriers flow (diffuse) across the junction. BJTs are commonly used in analog circuits, but they are also popular for use up to S-band for microwave applications [4].

Rather than create regions with electrons and holes, a similar effect can be attained by using multiple materials with different bandgaps called *heterojunctions*. An HBT can operate to well above 100 GHz. BJTs and HBTs made on silicon offer extremely low cost, high manufacturing yields, high gain, and low noise. Since silicon is a poor thermal conductor, BJTs and HBTs are generally impractical for high-power radars.

MOSFETs can deliver a lot of power and are still commonly used today to deliver 10s of amps into a load. They are generally limited to lower-frequency applications (L-band and below) due to high parasitic capacitance. LDMOS was developed to reduce the parasitic capacitance to enable operation to C-band [4].

MESFETs further improved the layout to enable operation to millimeter-wave (mm-wave). They were the transistor of choice for microwave applications, especially radar, for decades. Today, they are losing market share to HEMTs.

HEMT devices offer higher gain, lower noise, and higher operating frequency than MESFET at a competitive cost. Indium is commonly used as a heterojunction material because it improves high-frequency performance. GaN HEMTs have a heterojunction capable of supporting high power density. Since heterojunctions use materials with different bandgaps, they also have a different lattice size (spacing between atoms). This leads to discontinuities at the atomic layer that can "trap" charge, and this degrades device performance. Creating a smooth interface between the materials is a challenge, and there are two common approaches. A pseudomorphic HEMT (pHEMT) uses a very thin layer of one of the materials in the heterojunction, which allows the atomic structure to stretch. Alternatively, a metamorphic HEMT (or mHEMT) uses a buffer material in the middle of the heterojunction to fill in the gaps. Both approaches support higher-frequency operation.

Figure 3.6 shows a general small-signal (linear) model for a transistor. It is a three-port (or three-terminal) circuit with gate, drain, and source. Each port has a respective capacitance (C_g, C_d, C_s), inductance (L_g, L_d, L_s), and resistance (R_g, R_d, R_s). Additionally, each port has coupled parameters (C_{gs}, C_{ds}, C_{gd}, R_{gs}, R_{ds}, R_{gd}). There is also an input resistance (R_i) and a junction resistance (R_j). The voltage-controlled current source is related to the transconductance [g_m, defined in (3.20)] and the frequency (f) [4].

g_m can be calculated by:

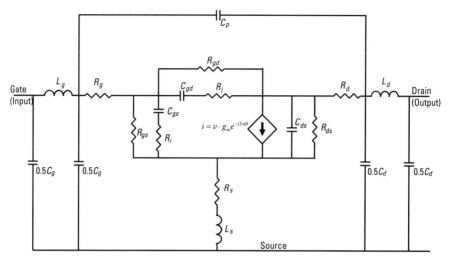

Figure 3.6 Equivalent small-signal circuit model of a real transistor.

$$g_m = \frac{I_d}{V_{gs}} = \frac{\Delta I_d}{\Delta V_{gs}} \tag{3.20}$$

where g_m (siemens, S) is the transconductance, I_d is the drain current (A), and V_{gs} is the gate-source voltage (volts, V). Sometimes g_m is written as the change in I_d divided by the change in V_{gs}.

g_m can also be calculated from:

$$g_m = \frac{2I_{MAX}}{|V_P|}\left(1 - \frac{V_{gs}}{V_P}\right) \tag{3.21}$$

where I_{MAX} is the maximum drain current (A), V_P is the pinch-off voltage (V), and V_{gs} is the gate-source voltage (V).

The transistor cutoff frequency (where it has unity gain) can be calculated from g_m and gate capacitance (C_{gd} is usually negligible):

$$f_T = \frac{g_m}{2\pi\left(C_{gs} + C_{gd}\right)} \approx \frac{1}{2\pi\tau_t} \tag{3.22}$$

where τ_t (seconds) is the transit time from the source to the drain. An approximation to the intrinsic delay of the transistor can be made from f_T.

The maximum gain possible [or *maximum available gain* (MAG)] from a transistor can be calculated from parameters within this model:

$$MAG = \left(\frac{f_T}{f}\right)^2 \frac{1}{4\dfrac{R}{R_{ds}} + 4\pi f_T C_{gd}\left(R + R_g + \pi f_T L_s\right)} \qquad (3.23)$$

where f_T is the cutoff frequency, f is the frequency of interest, R_{ds} is the drain-source resistance, C_{gd} is the gate-drain capacitance, R_g is the gate resistance, L_s is the source inductance, and R is the sum of $R_g + R_i + R_s + \pi f_T L_s$.

Practical Note

If the frequency doubles, (3.23) shows that MAG decreases by one-quarter or 6 dB. This is where the "–6 dB per octave" gain slope rule comes from.

The frequency where MAG is 1 is called f_{max} and it can be calculated from:

$$f_{max} = f_T \left[4\frac{R}{R_{ds}} + 4\pi f_T C_{gd}\left(R + R_g + \pi f_T L_s\right)\right]^{-1/2} \qquad (3.24)$$

From (3.22), (3.23), and (3.24), we can see that in order to operate at higher frequencies, transistor designers (usually device physicists) can do the following:

• Decrease C_{gs}, C_{gd};
• Decrease R_g, R_i, R_s;
• Decrease L_s.

These parameters can be modified by changing the channel doping concentration, gate width, channel thickness, gate length, source-gate-drain spacing, and the layer structure. Figure 3.7 shows an example of a GaAs HEMT cross section.

In a depletion-mode device, a conductive channel between the source and drain exists when 0V is applied to the gain (the device is nominally ON). The transistor is in the saturated region. As a negative voltage is applied to the gate, the transistor enters the linear region when the drain current varies with V_{gs} and V_{ds}. The transistor is essentially a voltage-controlled resistor (the word *transistor* is a contraction of *transfer resistor*). If the gate voltage is driven further negative,

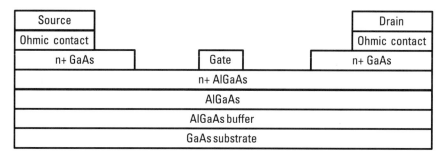

Figure 3.7   GaAs HEMT transistor cross section.

the channel will be shut off, and no drain current will flow. This is the cutoff region.

To map out the relationship between drain current (I_d), V_{gs}, and V_{ds}, an IV plot (current-voltage plot) can be made. Figure 3.8 provides an example of a IV plot [5].

The knee voltage (V_{knee}) marks the transition between the linear region (on the left) to the saturated region (on the right). The maximum current possible is denoted I_{MAX}. The saturated current level where $V_g = 0$ is called I_{DSS}. In the cutoff region, the drain current should become zero. Any current that continues to flow is leakage. In most cases, leakage is undesired and can be a sign of device failure. Some foundry processes have a small amount of drain current leakage, which is fine for amplifier applications, but the transistor cannot be used as a switch.

In an enhancement-mode device, the channel between the source and drain does not exist when 0V is applied to the gate (the device is nominally OFF). In this case, a positive voltage is applied to the gate to activate the chan-

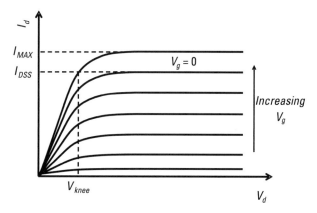

Figure 3.8   Example Id versus Vd plot (IV plot).

nel. The same terminology and IV curve characteristics as a depletion-mode device are also applicable for enhancement-mode devices.

3.3.3 Transistor Imperfections

Semiconductor wafers can be grown with extraordinary purity (measured in parts per billion). However, the layers that are deposited above the semiconductor will always have some defects. The epitaxial growth process responsible for adding the main transistor layers can introduce vacancies (missing atoms), interstitials (atoms in the wrong place), and impurities (undesired atoms) in the lattice structure. These defects can cause the following[3]:

- Current and voltage leakage;
- Deviation from normal behavior (generally from the trapping of impurities);
- Additional loss;
- Reduced frequency response;
- Catastrophic failure.

Foundries minimize defects during fabrication by carefully monitoring the following:

- Substrate and process temperature;
- Chamber pressure;
- Growth precursors;
- Wafer precleaning recipes.

Foundries tend to develop reputations for generating transistors with high reliability and low device-to-device variation. This is especially important in large AESA radars where uniformity across the structure is critical.

3.4 Custom Models

Models provided by component vendors should ideally be valid under all conditions (i.e., all substrates, frequencies, temperatures, and biases). Trying to represent the behavior accurately over the infinite trade space is impractical. Therefore, vendors optimize their models to be as broad reaching as possible. In doing so, compromise in advertised accuracy is made. For this reason, many designers choose to extract their own component models and optimize them for only the specific operating conditions required.

From measured data, a custom mathematical model can be created. Data that follows a line, parabola, or polynomial can be fitted to an equation using algebra.

It may be noticed that many of the mathematical models presented throughout this book leverage trigonometry (trig) or polynomial functions. When measurements are made of the natural world, the behavior tends to mimic a trend represented mathematically by a trig or polynomial function. Therefore, deriving a custom model starts by visually comparing the natural behavior against the library of functions, which are plotted in Figures 3.9–3.17 for convenience.

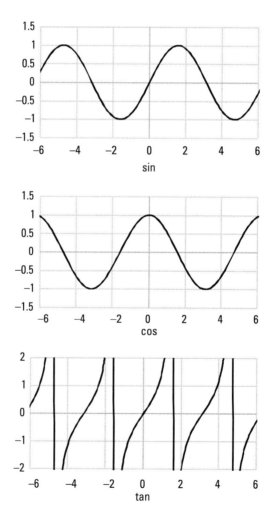

Figure 3.9   Plot of sine (sin), cosine (cos), and tangent (tan).

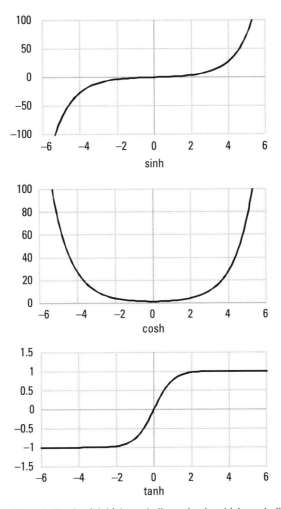

Figure 3.10 Plot of hyperbolic sine (sinh), hyperbolic cosine (cosh), hyperbolic tangent (tanh).

Once a suitable function is found, trial and error can be used to fit the model to the data.

3.5 Measurement Techniques

Using models to represent measured data offers design flexibility. Models can be scaled, tuned, and updated as needed. They offer superb resolution since parameters can be swept in small steps. Monte Caro analysis (discussed in Chapter 7) can be used to assess performance yield from tolerances.

However, if none of this flexibility is needed, modern simulators support the use of measurement files directly in the analysis. Various routines are

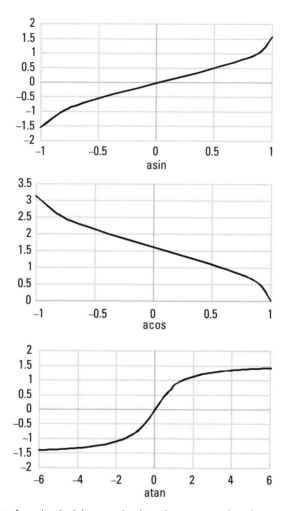

Figure 3.11 Plot of arcsine (asin), arccosine (acos), arctangent (atan).

available to extrapolate data or interpolate between steps. A measured S-parameter file (i.e., SnP file, where "n" is the number of ports) can be used in place of a small-signal (low-power) model. In place of a large-signal (high-power) model, load-pull measurements can be performed on a transistor to determine the ideal matching and bias conditions. Figure 3.18 shows a simple load-pull setup. Impedance tuners are placed at the input and output of the transistor. The tuners are controlled by a computer (not shown) to move around the Smith chart and record the input power, output power, and bias information. Once the

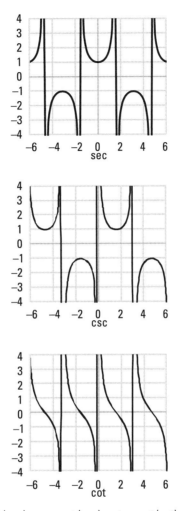

Figure 3.12 Plot of secant (sec), cosecant (csc), cotangent (cot).

impedance plane has been covered, the optimal gain/power or efficiency imped-ances are known. Rather than simulating with a device model, it is possible to design matching networks that present those measured impedances.

Many designers find this design approach too limiting since there is no flexibility in the bias (a measured data set is taken at one bias). However, for quick "down and dirty" designs, this approach is appropriate.

To assess stability, it is important to replace the transistor with an S2P (two-port S-parameter) file. This will provide the gain and match information needed (discussed in Chapter 4).

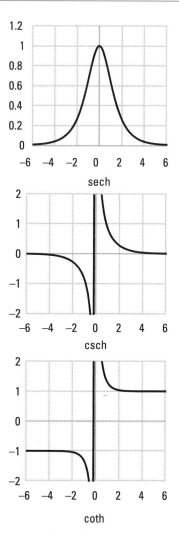

Figure 3.13   Plot of hyperbolic secant (sech), hyperbolic cosecant (csch), hyperbolic cotangent (coth).

Exercises

1. A designer places two inductors in series in order to achieve the desired inductance with parts available. Given that inductors have parasitic affects, is this placement beneficial or detrimental?

2. A designer places two capacitors in parallel in order to achieve the desired capacitance with parts available. Given that capacitors have parasitic affects, is this placement beneficial or detrimental?

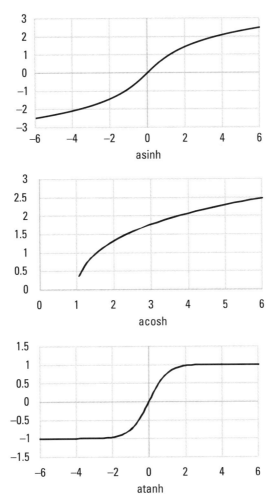

Figure 3.14 Plot of inverse hyperbolic sine (asinh), inverse hyperbolic cosine (acosh), inverse hyperbolic tangent (atanh).

3. Calculate the inductance of a 30-mil-long 28-gauge wire. How does that inductance change when placed 19 mils above a ground plane?

4. A series resonator circuit is comprised of a 10-pF capacitor, a 5-nH inductor, and a 10-Ω resistor. What are the bandwidth, quality factor, and equivalent impedance?

5. How would the behavior of the circuit from the previous exercise change if configured in parallel instead of in series?

6. Draw a true-scale IV plane for a transistor with $I_{DSS} = 1.5$A, $V_{knee} = 3$V, and $g_m = 4$S.

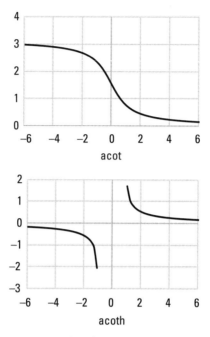

Figure 3.15   Plot of inverse cotangent (acot) and inverse hyperbolic cotangent (acoth).

7. The behavior of a component is best approximated as a hyperbolic cosecant for values greater than zero. Unfortunately, this trigonometric
function is not available on all modeling tools. What nontrigonomet-
ric function can be used as a substitute?

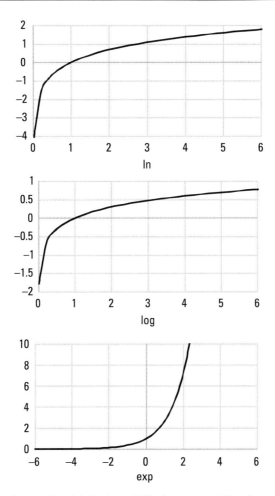

Figure 3.16 Plot of natural log (ln), log base 10 (log), exponential (exp).

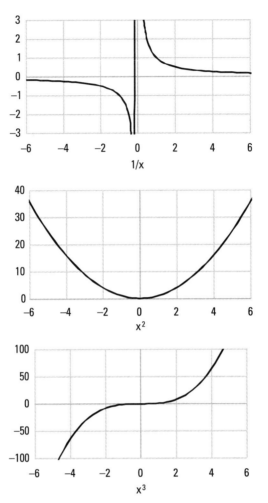

Figure 3.17 Plot of $1/x$, x^2, x^3.

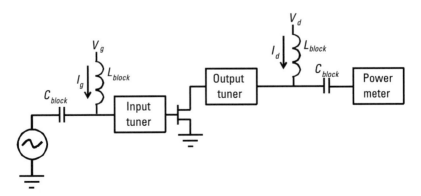

Figure 3.18 Simple load-pull setup.

References

[1]   Dorf, R., and J. Svoboda, *Introduction to Electric Circuits,* New York, NY: John Wiley & Sons, 1999.

[2]   Mantaro Product Development Services, Inc., "Impedance Calculators," Internet, http://www.mantaro.com/resources/impedance_calculator.htm.

[3]   Jaeger, R., *Microelectronic Circuit Design,* McGraw-Hill, 1997.

[4]   Senturia, S., *Microsystem Design*, Norwell, MA: Kluwer Academic Publishers, 1940.

[5]   Chang, K., I. Bahl, and V. Nair, *RF and Microwave Circuit and Component Design for Wireless Systems,* New York, NY: John Wiley & Sons, 2002.

Selected Bibliography

Maas, S., *Practical Microwave Circuits,* Norwood, MA: Artech House, 2014.

Wood, J., and D. Root, *Fundamentals of Nonlinear Behavioral Modeling for RF and Microwave Design,* Norwood, MA: Artech House, 2005.

Rudolph, M., *Introduction to Modeling HBTs,* Norwood, MA: Artech House, 2006.

Swanson, D., and W. Hoefer, *Microwave Circuit Modeling Using Electromagnetic Field Simulation,* Norwood, MA: Artech House, 2003.

Losee, F., *RF Systems, Components, and Circuits Handbook,* Norwood, MA: Artech House, 2005.

Part II
Component Design

4

Power Amplifier

One could argue that the power amplifier is the cornerstone of the active radar. Reviewing the equations for SNR_{search} and SNR_{track} in Chapter 1, the amplifier is the component that directly influences transmit power, system loss, noise figure, and frequency. It also consumes the most prime power and generates the most heat. It should come as no surprise that the bulk of new radar nonrecurring engineering (NRE) expense is often invested in the amplifier design.

One subject often overlooked by other amplifier design textbooks is the effect of mechanical constraints. As amplifiers achieve higher power in smaller volume, mitigating thermal effects becomes just as important as meeting electrical requirements. As thermal and environmental stress rise, so must the importance placed on mitigating mechanical effects. The best microwave component designers will have a basic understanding of mechanical engineering (and vice versa). For this reason, Chapter 4 presents both the electrical and mechanical aspects of power amplifier design for radar applications.

4.1 Amplifier Basics

Defining the specifications of an amplifier requires more metrics than just frequency, power, gain, efficiency, and linearity. Other important specifications include the following (listed in no particular order):

- Gain flatness over the operating band;
- Frequency roll-off characteristics out of band;
- Stability;
- VSWR (or return loss);

- Dynamic range;

- Prime power available (or power added efficiency, PAE);

- Linear phase response;

- Operating temperature;

- Duty cycle or peak-to-average power ratio (PAPR);

- Leakage current;

- Blanking requirement;

- Multifunctionality;

- Reconfigurability;

- Level of technology maturity;

- Sensitivity to production variation;

- Form factor;

- Bias available;

- Cost.

Practical Note

Unfortunately, specifications rarely include everything that a designer needs or wants to know. Or, specifications include the appropriate line items, but they are marked with "to be determined" or "TBD." This should not be confused with "not specified" or "NS." The former indicates there will be a specification, but one wasn't available at the time of printing. The latter indicates there will not be a specification.

There was once a specification generated for a high-power amplifier for a naval ship with an efficiency of "NS." When asked about it, the commander mentioned that he had twin 20-MW electric motors and an ocean full of cold water—efficiency wasn't a concern!

A schematic of the basic amplifier is shown in Figure 4.1. At the input is a voltage (RF) source, which represents the incoming signal. A capacitor (C_{block}) serves as a DC block to protect the RF source from over-current exposure. An input-matching network presents the ideal impedance to the transistor input. A gate voltage (V_g) is connected to the gate of a transistor through an inductor (L_{block}). The inductor serves as an RF block (or *RF choke*) to isolate the DC source from the amplifier. In some cases, L_{block} can be replaced with a large resistor or a combination of the two. On the transistor, the gate is at the input, the drain is at the output, and the source is grounded (known as a *common-source configuration*). The network at the output of the transistor is similar to that at the input. An output-matching network presents the ideal impedance to the

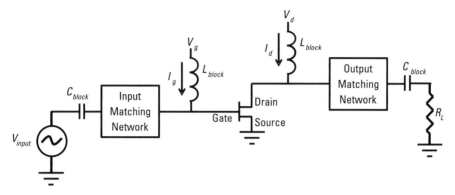

Figure 4.1   Schematic of basic amplifier.

transistor output. The drain voltage (V_d) is connected to the drain through an RF choke. A DC blocking capacitor protects the load (R_L) from excess DC bias.

If the input is single-tone (only one frequency present), then V_{input} can be represented as:

$$v_{in} = A\sin(2\pi f t + \phi) \tag{4.1}$$

where A is the amplitude (V), f is the frequency (Hz), t is the instance in time (s), and ϕ is the input phase (radians).

If the amplifier is perfectly linear with gain G, then the output can be represented as:

$$v_{out} = G \cdot A\sin(2\pi f t + \phi + \varphi) \tag{4.2}$$

where A is the amplitude (V), f is the frequency (Hz), t is the instance in time (s), φ is the input phase (radians), and φ is the amplifier phase shift (radians).

Expressed another way, the magnitude would be:

$$v_{out,rms} = \sqrt{2P_{out}Z_L} \tag{4.3}$$

where P_{out} is the output power (W) and Z_L is the load impedance (Ω).

In terms of the output current, from Ohm's law:

$$i_{out} = \frac{v_{out}}{Z_L} \tag{4.4}$$

where i_{out} is the output current (A), v_{out} is the output voltage (V), and Z_L is the load impedance (Ω).

In order to properly isolate the DC sources (V_g, V_d), an adequate inductance (L_{block}) must be chosen so the impedance "seen" looking into the DC source from the amplifier is very large. This can be calculated from:

$$Z_{block} = 2\pi f_{min} L_{block} \gg R \tag{4.5}$$

where L_{block} is the RF choke inductance (henrys), R is the desired resistance (Ω), and f_{min} is the maximum design frequency. R should be a large value. Choosing at least 20 times Z_o is usually adequate. For example, a 16-nH inductance would provide 1,000-Ω resistance at 10 GHz. For narrowband applications, L_{block} can be replaced with a quarter-wave transmission line.

Similarly, the values for the DC-blocking capacitors (C_{block}) must be carefully chosen to prevent unnecessary attenuation of the input RF signal. The resistance provided by the blocking capacitor can be calculated from:

$$Z_{C,block} = \frac{1}{2\pi f_{min} C_{block}} \ll R \tag{4.6}$$

where C_{block} is the DC blocking capacitance (F), R is the desired resistance (Ω), and f_{min} is the minimum design frequency. R should be a small value. Choosing 1/200 times Z_o is usually adequate. For example, a 64-pF capacitor would provide 0.25-Ω resistance at 10 GHz.

Amplifiers are often categorized into classes to help convey the type of design technique used. In conversation, the mere mention of a class can give an impression to the listener of output power, efficiency, linearity, frequency limitations, and SWAP-C. The most popular classes for RF amplifiers are A, B, AB, C, E, and F, as described in Sections 4.1.1–4.1.6.

4.1.1 Class A

Class A is the most linear of the classes. It is often used as a gain stage since it offers high gain and doesn't degrade the linearity and noise figure of the later stages. Figure 4.2 shows the operation [1]. Recall that the gate is the transistor input and that the drain is the transistor output.

The current I_{DQ} is biased at 50% of I_{MAX}, which puts it along the middle of the IV plane. The voltage V_{DQ} is set near the center of the IV plane and should be less than half the breakdown voltage ($< V_{BR}/2$) of the transistor. The load line is drawn on the I_d-V_d plot from the knee voltage on the highest current trace through the normal operating bias (called the Q-point) to the X-axis. During operation, the voltage moves (*swings*) along the load line. The decreasing voltage ($V_{swing,-}$) will swing between V_{knee} and V_{DQ}. Just like a pendulum, the increasing voltage ($V_{swing,+}$) will have the same magnitude as $V_{swing,-}$, but in the

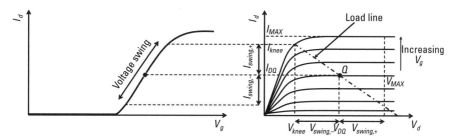

Figure 4.2 Class A Id versus V_g and I_d versus V_d plots.

opposite direction. The maximum drain voltage is labeled V_{MAX}. The peak-to-peak voltage (V_{pp}) is calculated from $V_{MAX} - V_{knee}$ or $2V_{swing,-}$.

A similar behavior happens with the current swing. The increasing current $I_{swing,+}$ will swing between I_{knee} and I_{DQ}. The decreasing current $I_{swing,-}$ will have the same magnitude as $I_{swing,+}$, but in the opposite direction. Since I_{DQ} is set to 50% I_{MAX}, the current swing should never need to go below 0 A.

The voltage and current swing can simultaneously be plotted on the transconductance (I_d-V_g) plot. As shown, the swing is wholly contained within the linear region of the transconductance curve. This gives a Class A amplifier unique qualities, described as follows:

- A Class A amplifier operates in a linear mode, by definition.
- The input power level must not exceed $P_{max}/2$, where P_{max} is defined in (4.7). This will allow a class A amplifier to continuously conduct (100% of the voltage swing is above 0A so it always draws power). Said another way, it has a conduction angle of 360° (it is always conducting).
- Since Class A is always "on," there is no "turn on" time required.
- The theoretical maximum efficiency is 50%.

The maximum output power (W) is calculated from:

$$P_{\max} = \frac{V_{pp}I_{pp}}{8} = \frac{(V_{swing,-} + V_{swing,+})(I_{swing,-} + I_{swing,+})}{8} \qquad (4.7)$$

where V_{pp} is the peak-to-peak voltage (V) and I_{pp} is the peak-to-peak current (A).

To increase P_{max}, V_{DQ} should also be maximized as long as V_{MAX} is less than the breakdown voltage of the transistor. Damage will occur to the transistor if the operating voltage exceeds breakdown.

For maximum power, the load impedance should be set to:

$$Z_L = \frac{V_{pp}}{I_{pp}} = \frac{(V_{swing,-} + V_{swing,+})}{(I_{swing,-} + I_{swing,+})} \qquad (4.8)$$

4.1.2 Class B

Class B is not as linear as Class A, but it is significantly more efficient (theoretical max is 78.5%). It is often used in a push-pull configuration (explained in Section 4.3.3). The operation is shown in Figure 4.3 [1].

The current I_{DQ} is biased at pinch-off, which puts it along the bottom of the IV plane. As with Class A, the voltage V_{DQ} is set near the center of the IV plane and should be less than half the breakdown voltage ($< V_{BR}/2$) of the transistor. The load line is drawn on the I_d-V_d plot from the knee voltage on the highest current trace through the Q-point to the X-axis. During operation, the voltage swings in the same manner as Class A.

The current behavior is quite different. The increasing current $I_{swing,+}$ will still swing between I_{knee} and I_{DQ}. However, like a pendulum, the decreasing current $I_{swing,-}$ wants to swing negative, but I_d does not go negative—it stays at 0 A. This is best seen when plotted on the transconductance (defined in Chapter 3) plot. Half the time, current is conducting and half the time the current is 0 A.

This gives a Class B amplifier unique qualities, described as follows:

- Since the voltage swing is nonlinear, an amplifier operating Class B will also be nonlinear.

- A Class B amplifier is conducting half the time (50% of the voltage swing is above 0A). Said another way, it has a conduction angle of 180°.

- Since conducting angle is less than 360°, gain will be less than Class A (it must be driven with more input power to achieve maximum output power).

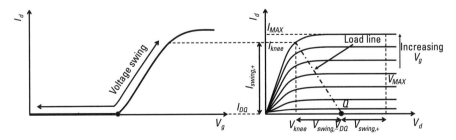

Figure 4.3   Class B Id versus V_g and Id versus V_d plots.

- The theoretical maximum efficiency is $\pi/4 \approx 78.5\%$.

The maximum output power (W) is calculated from:

$$P_{max} = \frac{V_{pp}I_{pp}}{8} = \frac{(V_{swing,-} + V_{swing,+})(I_{swing,+})}{8} \tag{4.9}$$

where V_{pp} is the peak-to-peak voltage (V) and I_{pp} is the peak-to-peak current (A).

As with Class A, to increase P_{max}, V_{DQ} should be maximized as long as V_{MAX} is less than the breakdown voltage of the transistor.

The load impedance should be set to:

$$Z_L = \frac{V_{pp}}{I_{pp}} = \frac{(V_{swing,-} + V_{swing,+})}{I_{swing,+}} \tag{4.10}$$

Transistors have both forward and reverse breakdown voltages. As mentioned previously, V_{DQ} must be chosen to ensure V_{MAX} does not exceed the breakdown voltage (with some margin). On the transconductance plot, if the voltage swings too far to the left (very negative V_g), I_d will begin to conduct again. Designers must also ensure that V_g stays above the reverse breakdown voltage of the transistor.

To maximize efficiency, Class B amplifiers can shape the voltage waveforms to decrease the dissipated power within the transistor. This can be accomplished by terminating all harmonics with a short circuit (SC). In Figure 4.4, a parallel shunt LC circuit has been added to provide a SC at $2f_o$ and $3f_o$. This helps shape the output waveform into a sinusoid.

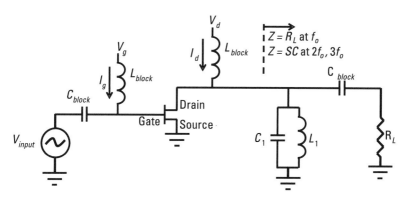

Figure 4.4   Example Class B schematic.

4.1.3   Class AB

Class AB strives to compromise between the linearity of Class A and the efficiency of Class B. It is arguably the most commonly used configuration. The operation is shown in Figure 4.5 [1].

The current (I_{DQ}) is biased between pinch-off and 50% of I_{MAX}. Often, it is set to 25% of I_{MAX}, but can range between 5 and 30%. As with Class A and Class B, the voltage (V_{DQ}) is set near the center of the IV plane and should be less than half the breakdown voltage ($< V_{BR}/2$) of the transistor. The load line is drawn on the I_d-V_d plot from the knee voltage on the highest current trace through the Q-point to the X-axis. During operation, the voltage swings in the same manner as class A and class B.

The increasing current ($I_{swing,+}$) will still swing between I_{knee} and I_{DQ}. In this case, the decreasing current ($I_{swing,-}$) cannot make a full swing (like Class A) but it can swing more than Class B. This is best seen when plotted on the transconductance plot. More than half the time, current is conducting, but part of the swing is at 0A.

This gives a Class AB amplifier unique qualities, listed as follows:

- Linearity is between Class A and Class B;
- Efficiency is between Class A and Class B (between 50 and 78.5%);
- Gain is between Class A and Class B;
- It has a conduction angle between 180° and 360°.

The maximum output power (W) is calculated from:

$$P_{max} = \frac{V_{pp}I_{pp}}{8} = \frac{(V_{swing,-} + V_{swing,+})(I_{swing,-} + I_{swing,+})}{8} \qquad (4.11)$$

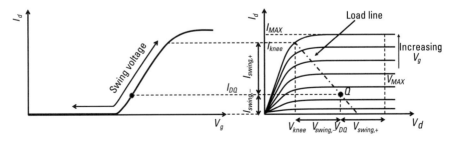

Figure 4.5   Class AB I_d versus V_g and I_d versus V_d plots.

where V_{pp} is the peak-to-peak voltage (V) and I_{pp} is the peak-to-peak current (A).

The load impedance should be set to:

$$Z_L = \frac{V_{pp}}{I_{pp}} = \frac{(V_{swing,-} + V_{swing,+})}{(I_{swing,-} + I_{swing,+})} \quad (4.12)$$

To maximize efficiency, Class AB amplifiers can shape the voltage waveforms to decrease the dissipated power within the transistor. This can be accomplished by terminating all harmonics with an SC. The same Class B schematic applies to Class AB.

4.1.4 Class C

Class C is operated well below pinch-off. This gives it the best efficiency and the worst linearity of the classes presented so far. Due to its nonlinear nature, it is limited to frequency- or phase-modulated waveforms or applications with a liberal linearity requirement. The operation is shown in Figure 4.6 [1].

It is biased deep into pinch-off so without a large-signal RF signal applied, it is off. In comparison with the other classes, only a small positive current ($I_{swing,+}$) swing is achieved under input drive. Class C also benefits from harmonic matching, like Class B. This gives a Class C amplifier unique qualities, outlined as follows:

- Linearity is worse than Class B;
- Efficiency is better than Class B (up to 81%);
- Gain is less than Class B;
- It has a conduction angle less than 180° (typically 90–120°).

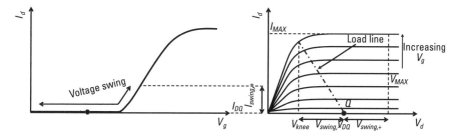

Figure 4.6 Class C I_d versus V_g and I_d versus V_d plots.

Practical Note

Measuring S-parameters on a Class C amplifier will not show gain (S_{21}), which can cause great confusion during measurements. Input power must be driven to draw current and see RF gain.

4.1.5 Harmonically Matched Classes

Class AB, Class B, and Class C can benefit from harmonically matched outputs. Providing an SC impedance at the harmonics shapes the voltage waveforms within the transistor to reduce current and voltage overlap and reduce the dissipated power ($P_{diss} = I_d \cdot V_d$). Some classes use this principle to attempt to eliminate all current and voltage overlap to approach a theoretical 100% efficiency.

Class E is the first of several topologies with a theoretical 100% efficiency. The transistor is terminated with a high-Q circuit to provide a reactive load at f_o and open circuit at $2f_o$ and $3f_o$. A perfect implementation causes the transistor to switch on and off, giving it very precise waveforms (hence the term *switch-mode amplifier*). This precision minimizes voltage/current overlap and provides very high efficiency. Unfortunately, realizing a Class E amplifier in practice is very challenging and limiting. The high-Q circuit limits the operating frequency range to very narrow bandwidth. The switching speed must be fast compared with the frequency, which limits operation to low frequencies (L-band or below principally). For these reasons, Class E is not widely used.

Class F (and its inverse, Class F^{-1}) is gaining in popularity because it works well at microwave frequencies. The ideal Class F will have all odd harmonics terminated in an open circuit and all even harmonics terminated in a short circuit. Class F^{-1} terminates all odd harmonics with a short circuit and all even harmonics with an open circuit. Ideally, both input and output will follow this termination scheme, although in practice the focus is on the output. Figure 4.7 shows a circuit approach for Class F [1].

In the circuit shown in Figure 4.7, L_1 and C_1 resonate at f_o, L_3 and C_3 resonate at $3f_o$, and C_2 provides blocking and compensation for parasitics (if any). These values can be calculated by [1]:

$$C_1 = \frac{1 - 0.5\dfrac{BW}{f_o}}{2\pi f_o Z_L\left(1 - \left[1 - 0.5\dfrac{BW}{f_o}\right]^2\right)} \tag{4.13}$$

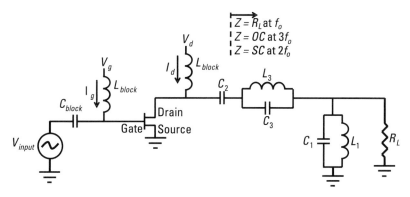

Figure 4.7 Example class F schematic.

where C_1 is the capacitance (F), BW is the design fractional bandwidth (Hz), f_o is the design frequency (Hz), and Z_L is the load impedance (Ω).

$$L_1 = \frac{1}{4\pi^2 f_o^2 C_1} \tag{4.14}$$

where L_1 is the inductance (H), f_o is the design frequency (Hz), and C_1 is the capacitance previously calculated (F).

$$L_3 = 1.9753 \frac{L_1 R_L^2}{9R_L^2 + 16\pi^2 f_o^2 L_1^2} \tag{4.15}$$

where L_3 is the inductance (H), L_1 is the previously calculated inductance (H), R_L is the load resistance (Ω), and f_o is the design frequency (Hz).

$$C_3 = \frac{1}{36\pi^2 f_o^2 L_3} \tag{4.16}$$

where C_3 is the capacitance (F), L_3 is the previously calculated inductance (H), and f_o is the design frequency (Hz). R_L is calculated in the same way as Class B.

Class F can also be implemented using distributed instead of lumped elements. In either implementation, two factors, detailed in the following, make 100% efficiency unobtainable in practice.

- Individual elements (i.e., capacitors, inductors, and transmission lines) have parasitics, which prevent achieving perfect open and SCs.

- Terminating all harmonics would require an infinite number of elements (which in itself is impractical). In addition to parasitics, real elements have loss. Eventually, the benefit of terminating additional harmonics is outweighed by the loss of adding more elements.

For these reasons, most implementations do not handle input harmonics and only terminate $2f_o$ and $3f_o$ on the output. Some will terminate $2f_o$ on the input and $4f_o$ on the output.

Practical Note

Although achieving Class F over very broad bandwidth is not practical, it is generally suitable for radar bandwidths (up to 1 GHz).

4.1.6   Do Classes Really Matter?

Understanding how classes work is an important step in learning amplifier design. It will help gain an intuition between the balance of bias, impedance termination, output power (or gain), efficiency, and linearity. In conversation, if someone says he or she is targeting Class AB operation, it is universally known throughout the amplifier design community what that means.

That said, when it comes to designing amplifiers, the answer to Section 4.1.6's title question is (ultimately) no. Specifications almost never have a specified class of operation. Instead, they focus on factors such as power, efficiency, and linearity (as listed in Section 4.1). It's up to the designer to determine the ideal bias condition and whether harmonic terminations are worthwhile. Even if one wanted to realize a perfect textbook implementation of a particular class, there is no way to verify it. True verification would require measuring the precise voltage and current at the transistor terminals (not the amplifier terminals), which is impractical due to the *observer effect* (the act of probing would alter the waveforms they were intended to measure).

Some of the best amplifiers made in practice do not follow any of the before mentioned classes. A combination of clever matching, bias, and transistor selection can shape waveforms in such a way to support high conduction angles (for improved gain, power, and linearity) and reduce current/voltage overlap (for improved efficiency). Unfortunately, the best design strategies are often kept as trade secrets so they cannot be discussed openly.

4.2   Design Strategies and Practices

Understanding how bias and terminations effect amplifier behavior allows a designer to choose the best approach to fit a particular performance need. Once that is established, the details must be fine-tuned to ensure all requirements are met. This section dives deeper into amplifier design details.

4.2.1 Stability

All active devices have the potential of becoming unstable. If this happens, power is generated at a frequency other than what is intended. This undesired frequency is called an *oscillation*, and the amplifier is said to be oscillating when this happens. Since energy must be conserved, this means less energy is generated at the frequencies of interest (i.e., f_o). Aside from a decrease in performance, instability can also lead to failure if driven harder or operated long enough.

There are many potential sources of oscillation. Often, stability is linked to certain input or output impedances. An amplifier that is stable regardless of the port impedances is *unconditionally stable*. It is almost always desirable to be unconditionally stable (if the port impedances are tightly controlled, it is conceivable that unconditional stability may not be required).

Fortunately, there are ways during the design process to determine if a circuit can be unstable and to resolve the source of instability. The methods described in this section must be true over all frequencies, not just the intended operating frequency. Instability can occur well below or above band. When performing a stability analysis, it is best to calculate or simulate from 0 Hz (or as close to it as possible) to f_{max} (3.24). At minimum, analysis should be performed to at least $2f_o$ ($3f_o$ or more would be better).

The quickest test is to evaluate the return loss. If S_{11} and S_{22} are not contained within the Smith chart, then the amplifier is unstable. That is, to be stable,

$$|S_{11}| < 1 \tag{4.17}$$

and

$$|S_{22}| < 1 \tag{4.18}$$

A more thorough analysis includes calculating the *Rollett stability factor*, or *K-factor*, which must be greater than one for unconditional stability . Making this calculation starts by calculating the determinant of the S-parameters (denoted as Δ).

$$\Delta = S_{11}S_{22} - S_{12}S_{21} \tag{4.19}$$

The K-factor equation is [1]:

$$K_{factor} = \frac{1 - |S_{11}|^2 - |S_{22}|^2 + |\Delta|^2}{2|S_{12}S_{21}|} \qquad (4.20)$$

A second analysis includes calculating the *stability measure*, which must be greater than zero [2]:

$$K_{measure} = 1 + |S_{11}|^2 - |S_{22}|^2 - |\Delta|^2 \qquad (4.21)$$

Graphically, stability circles can be plotted on a Smith chart to show impedance regions of stability (or instability). Separate circles are drawn for the source and load. The centers of the circles ($C_{stability,L}$ and $C_{stability,S}$) are calculated from [1]:

$$C_{stability,L} = \frac{S_{11}\Delta^* - S_{22}^*}{|\Delta^2| - |S_{22}|^2} \qquad (4.22)$$

$$C_{stability,S} = \frac{S_{22}\Delta^* - S_{11}^*}{|\Delta^2| - |S_{11}|^2} \qquad (4.23)$$

where "*" indicates the complex conjugate of the number should be taken (flip the sign on the imaginary part). So, the complex conjugate of $0.25 - j0.5$ would be $0.25 + j0.5$.

The radii of the circles ($R_{stability,L}$ and $R_{stability,S}$) are calculated from [1]:

$$R_{stability,L} = \frac{|S_{12}S_{21}|}{\left|\left(|\Delta|^2 - |S_{22}|^2\right)\right|} \qquad (4.24)$$

$$R_{stability,S} = \frac{|S_{12}S_{21}|}{\left|\left(|\Delta|^2 - |S_{11}|^2\right)\right|} \qquad (4.25)$$

From the center and radius, a compass can be used to draw the circles on a Smith chart. The circle drawn can either enclose the region of stability or instability. This is determined from and whether the origin ($Z_n = 1 + j0\Omega$) is within the circle or outside the circle. Table 4.1 summarizes these conditions.

Alternatively, the same conclusion made in Table 4.1 can be determined mathematically. The load stability circle encloses the origin if either are true:

Table 4.1
Conditions for Determining Type of Stability Circle

$|S_{11}|$	Origin Within Circle	Origin Outside Circle
<1	Interior stable	Exterior stable
>1	Exterior stable	Interior stable

$$|S_{11}| < 1 \text{ and } |\Delta| > |S_{22}| \qquad (4.26)$$

$$|S_{11}| > 1 \text{ and } |\Delta| < |S_{22}| \qquad (4.27)$$

The source stability circle encloses the origin if either are true:

$$|S_{22}| < 1 \text{ and } |\Delta| > |S_{11}| \qquad (4.28)$$

$$|S_{22}| > 1 \text{ and } |\Delta| < |S_{11}| \qquad (4.29)$$

Another way to determine stability is to calculate the geometrically derived stability factor (μ) [2]. If the quantity is greater than one, the circuit is stable.

$$\mu_{load} = \frac{1 - |S_{11}|^2}{|S_{22} - S_{11}^* \cdot \Delta| + S_{12} \cdot S_{21}} \qquad (4.30)$$

$$\mu_{source} = \frac{1 - |S_{22}|^2}{|S_{11} - S_{22}^* \cdot \Delta| + S_{12} \cdot S_{21}} \qquad (4.31)$$

Chapter 8 discusses methods for correcting instability.

4.2.2 Power and Gain

The output power capability of a transistor is proportional to its periphery (more specifically, its gate periphery). The periphery is equal to the gate width multiplied by the number of gate fingers). A $4 \times 200\text{-}\mu\text{m}$ transistor has a total periphery of 800 μm (0.8 mm). If the power density of the transistor is 3.5 W/mm, then the maximum output power of the transistor is 3.5 W/mm $\times$ 0.8 mm = 2.8W. If the desired transistor output power is 25W, then nine of those $4\times200\text{-}\mu\text{m}$ transistors are needed.

Output power at the amplifier level is often specified at a saturated level (P_{sat}, meaning no additional power is possible regardless of increasing input drive) or at a particular compression level. On a spec sheet, the latter appears as PNdB, where N is the number of decibels compressed (i.e., gain has reduced by N dB from the small-signal level). P1dB is commonly used and denotes 1-dB gain compression. P3dB is also commonly used since it is often similar in power to P_{sat} and is more precisely defined.

Practical Note

P1dB is a common power level because it generally marks the compression point when linearity starts to degrade quickly. However, we know from our discussion on classes that some amplifiers will be inherently more linear than others. It is much more accurate to define "linear output power," which is the power level required without exceeding a set linearity. For example, a specification could include an IM3 specification of 40 dBc (decibels relative to the carrier) with a linear output power of 10W. From a system point-of-view, the compression level is not important. The output power and linearity are what matter.

Transistor impedance is inversely proportional to periphery. The larger a transistor is scaled, the smaller the impedance becomes and the more difficult it becomes to match to (especially broadband). For high-power applications, using transistors with higher power density (like GaN) is advantageous because smaller transistors can be used.

The effects of impedance mismatch were discussed in Chapter 2 and now will be applied to power amplifiers. When an amplifier is attached to a signal source, a potential mismatch occurs. If the effects of the mismatch are ignored, the gain of the system is called the *power gain*, and it is defined by:

$$G_p = \frac{P_L}{P_{in}} \qquad (4.32)$$

where G_p is the power gain (unitless), P_L is the power delivered to the load (watts), and P_{in} is the amplifier input power (watts).

If the effects of the mismatch are not ignored, the gain of the system is called the *transducer gain* and it is defined by:

$$G_t = \frac{P_L}{P_{inc}} \qquad (4.33)$$

where G_t is the transducer gain (unitless), P_L is the power delivered to the load (watts), and P_{inc} is the available source input power or incidence power (watts).

The input power is related to the incident power by:

$$P_{in} = P_{inc}\left(1 - |\Gamma|^2\right) \tag{4.34}$$

where Γ is the input reflection coefficient (unitless).

The power and transducer gain can also be calculated from S-parameters [1]:

$$G_p = \frac{\left(1 - |\Gamma_L|^2\right)|S_{21}|^2}{|1 - S_{22}\Gamma_L|^2 - |S_{11} - \Delta\Gamma_L|^2} \tag{4.35}$$

$$G_t = \frac{\left(1 - |\Gamma_L|^2\right)\left(1 - |\Gamma_S|^2\right)|S_{21}|^2}{|1 - S_{22}\Gamma_L - S_{11}\Gamma_S + \Delta\Gamma_S\Gamma_L|^2} \tag{4.36}$$

where Γ_L is the load reflection coefficient, Γ_S is the source reflection coefficient, and Δ is the S-parameter determinant (4.19).

When the circuit is conjugate matched at the load, maximum gain is achieved and the following condition is met:

$$G_p = G_t = G_{max} = \left|\frac{S_{21}}{S_{12}}\right|\left(K - \sqrt{K^2 - 1}\right) \tag{4.37}$$

where K is the Rollett's Stability Factor (4.20).

The maximum stable gain is achieved when $K = 1$ and G_{max} becomes G_{msg}:

$$G_{msg} = \left|\frac{S_{21}}{S_{12}}\right| \tag{4.38}$$

Just as with stability circles, gain circles can also be drawn. This allows a designer to see graphically the trade space with other performance metrics. Since everything plotted on a Smith chart is normalized, graphing gain circles starts by calculating the maximum normalized gain [3]:

$$G_{n,max} = \frac{G_{max}}{|S_{21}|^2} = \frac{\left|\frac{S_{21}}{S_{12}}\right|\left(K - \sqrt{K^2 - 1}\right)}{|S_{21}|^2} = \frac{K - \sqrt{K^2 - 1}}{|S_{12}S_{21}|} \tag{4.39}$$

Then, a desired gain level to plot can be calculated from:

$$G_n = \frac{G}{|S_{21}|^2} \tag{4.40}$$

where G is the desired gain level (unitless). G_n must be less than $G_{n,max}$.
The center of the gain circle is located at [1]:

$$C_{gain,L} = \frac{G_n\left(S_{22}^* - \Delta^* S_{11}\right)}{G_n\left(|S_{22}|^2 - |\Delta|^2\right) + 1} \tag{4.41}$$

The radius of the gain circle is [1]:

$$R_{gain,L} = \frac{\sqrt{1 - 2KG_n|S_{12}S_{21}| + G_n^2|S_{12}S_{21}|^2}}{\left|G_n\left(|S_{22}|^2 - |\Delta|^2\right) + 1\right|} \tag{4.42}$$

This process can be repeated to get a set of gain contours.

4.2.3 Efficiency

Providing good efficiency is important for a number of reasons at the system and component levels. The following are system-level reasons for efficiency:

- Reducing prime power consumption while maintaining desired output performance;
- Producing prime power with a smaller generator;
- Generating less waste heat and thereby minimizing thermal load;
- Lowering operating expense;
- Improving mobility due to smaller size and weight.

The component-level benefits of efficiency are the following.

- Longer lifetime since the component;
- The lower bias (mainly drain current) simplifies circuitry.

The most common efficiency metric is the PAE, which is calculated from:

$$PAE(\%) = \frac{P_L - P_{in}}{P_{dc}} \times 100 = \frac{P_L}{P_{dc}}\left(1 - \frac{1}{G_p}\right) \times 100 = \eta_d \left(1 - \frac{1}{G_p}\right) \times 100 \quad (4.43)$$

where P_L is the power at the load (W), P_{in} is the input power (W), P_{dc} is the DC power consumed (W), G_p is the power gain (unitless), and η_d is the drain efficiency (unitless).

The DC power consumed (P_{dc}) can be calculated from:

$$P_{dc} = \sum_{i=1}^{N}\left(V_{d,i}I_{d,i} + V_{g,i}I_{g,i}\right) \quad (4.44)$$

Often I_g is negligible compared to I_d, so that term is omitted. Drain efficiency is another efficiency metric that does not subtract input power. Drain efficiency can be calculated from:

$$\eta_d(\%) = \frac{P_L}{P_{dc}} \times 100 \quad (4.45)$$

When gain is high, P_L is much larger than P_{in} so η_d is similar to PAE.

4.2.4 Gain Flattening

It was shown in (3.23) that the gain drops by 6 dB every time the frequency doubles. In some radars, this is not an issue. The radar range will be larger at lower frequencies. In some radars, gain flatness over the band is critical. There are two principal methods for flattening gain, described as follows.

- Adding a lossy filter like the pi-network shown in Figure 4.8 to the input of the amplifier to equalize the gain. Although the circuit can be placed anywhere in the RF chain, it is usually placed right before the amplifier. Values can be optimized to track any unique characteristics of the gain roll-off. There are a myriad of high-pass filters that can be used for this approach.

- Adding a negative feedback circuit between the gate and drain. This is commonly implemented with a resistor as shown in Figure 4.9. To prevent shorting V_g and V_d, a large capacitor is placed in series with the resistor. Feedback mechanisms are complicated, so in practice, the value of R is generally determined through optimization in a simulator.

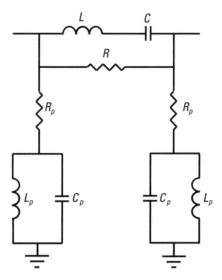

Figure 4.8 Example of a gain-flattening circuit.

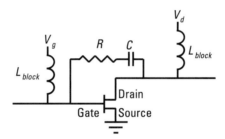

Figure 4.9 Example negative-feedback circuit.

Adding negative feedback also tends to improve the input and output return loss and improve stability. However, it can be tricky to implement since the RC circuit could be physically much larger than the transistor (especially if in bare die form). To be effective, the RC circuit needs to be as close to the gate and drain as possible to minimize phase shift in the feedback loop. Since a resistor is added to the series path, noise figure will degrade.

4.2.5 VSWR

Chapter 2 discusses the concepts of VSWR and mismatch. For an amplifier, achieving good VSWR is important to ensure maximum incident power and maximum delivered power to the load. The reflection coefficient for the input and output can be calculated from:

$$\Gamma_{in} = \frac{\Delta\Gamma_L - S_{11}}{S_{22}\Gamma_L - 1} \tag{4.46}$$

$$\Gamma_{out} = \frac{\Delta\Gamma_S - S_{22}}{S_{11}\Gamma_S - 1} \tag{4.47}$$

where Δ is the S-parameter determinant (4.19), Γ_L is the load reflection coefficient, and Γ_S is the source reflection coefficient.

The degree of mismatch for a given gamma of interest, Γ_x, can be computed by:

$$M_{in} = \frac{\left(1 - |\Gamma_{in}|^2\right)\left(1 - |\Gamma_x|^2\right)}{|1 - \Gamma_{in}\Gamma_x|^2} \tag{4.48}$$

$$M_{out} = \frac{\left(1 - |\Gamma_{out}|^2\right)\left(1 - |\Gamma_x|^2\right)}{|1 - \Gamma_{out}\Gamma_x|^2} \tag{4.49}$$

From this information, constant source and load VSWR circles can be created. The centers of those circles are located on a Smith chart at:

$$C_{match,S} = \frac{M_{in}\Gamma_{in}^*}{1 - \left(1 - M_{in}\right)|\Gamma_{in}|^2} \tag{4.50}$$

$$C_{match,L} = \frac{M_{out}\Gamma_{out}^*}{1 - \left(1 - M_{out}\right)|\Gamma_{out}|^2} \tag{4.51}$$

The radius of those circles can be calculated from:

$$R_{match,S} = \frac{\sqrt{1 - M_{in}}\left(1 - |\Gamma_{in}|^2\right)}{1 - \left(1 - M_{in}\right)|\Gamma_{in}|^2} \tag{4.52}$$

$$R_{match,L} = \frac{\sqrt{1 - M_{out}}\left(1 - |\Gamma_{out}|^2\right)}{1 - \left(1 - M_{out}\right)|\Gamma_{out}|^2} \tag{4.53}$$

This process can be repeated for different Γ_x values to generate a set of contours. These contours can be plotted on the same Smith chart as gain and stability circles to see the trade space graphically.

4.2.6 Conjugate Matching

If the transistor is unconditionally stable, conjugate impedance matching can be used to achieve peak gain and power. This is implemented using the following steps.

- The conjugate of the input impedance should be set to the source impedance.

$$Z_s = Z_{in}^*$$ (4.54)

- The conjugate of the input reflection coefficient should be set to the source reflection coefficient.

$$\Gamma_s = \Gamma_{in}^*$$ (4.55)

- The conjugate of the output impedance should be set to the load impedance.

$$Z_L = Z_{out}^*$$ (4.56)

- The conjugate of the output reflection coefficient should be set to the load reflection coefficient.

$$\Gamma_L = \Gamma_{out}^*$$ (4.57)

- The source and load reflection coefficients should be set to:

$$\Gamma_S = \frac{1}{2B_1}\left[A_1 \pm \sqrt{A_1^2 - 4|B_1|^2} \right]$$ (4.58)

$$\Gamma_L = \frac{1}{2B_2}\left[A_2 \pm \sqrt{A_2^2 - 4|B_2|^2} \right]$$ (4.59)

where a minus sign is used when $A_n > 0$, a plus sign is used when $A_n < 0$, and:

$$A_1 = 1 + |S_{11}|^2 - |S_{22}|^2 - |\Delta|^2 \tag{4.60}$$

$$A_2 = 1 + |S_{22}|^2 - |S_{11}|^2 - |\Delta|^2 \tag{4.61}$$

$$B_1 = S_{11} - \Delta S_{22}^* \tag{4.62}$$

$$B_2 = S_{22} - \Delta S_{11}^* \tag{4.63}$$

4.2.7 DC Bias Filtering

Any alternating current (AC) or RF signal that is superimposed on the DC line will enter the transistor drain and be amplified. In addition to unwanted (spurious) signals on the output, this can also lead to oscillation. DC lines can pick up AC or RF signals by:

- Noise from the power supply (even the best models have a noise specification);
- Coupling from other circuit elements;
- Receiving signals from the environment outside the amplifier.

These non-DC signal components can be removed by providing a low resistance path to ground without shorting DC to ground. This is accomplished using a decoupling or bypass capacitor. If the component value is chosen properly, any alternating voltage passing that node should be shorted to ground. Since a capacitor blocks DC, the DC voltage is unaffected.

The best decoupling circuits utilize multiple capacitors in parallel rather than one large capacitor. (Remember, larger capacitors provide low resistance to lower frequencies.) This provides a balance of component response time and parasitic mitigation. A good rule of thumb is to use three or more parallel capacitors. Figure 4.10 shows an amplifier with gate and drain decoupling circuits.

Since noise can be comprised of virtually any frequency, a wideband response is needed. Generally, small, medium, and large capacitors are chosen to satisfy this need. For example, the largest capacitor may be 1–10 μF (although it is not uncommon to go as high as 220–470 μF). The medium capacitor may be between 0.01 and 0.1 μF. The small capacitor is usually 100–1,000 pF.

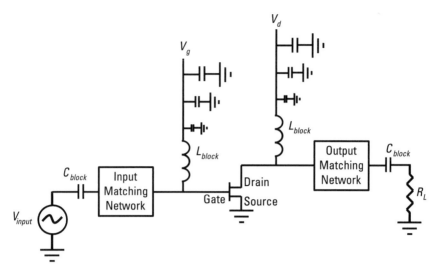

Figure 4.10   Circuit with proper DC bias filtering.

Decoupling capacitors should be added wherever there are long DC (or bias) transmission lines. If placed every $\lambda/8$ or so, this should provide adequate protection. Chapter 7 discusses decoupling circuits for pulsed applications.

4.2.8   Multistage Amplifiers

Depending on frequency, technology, and class of operation, the gain from a single transistor can range from 10 to 20 dB. Generally, this is not adequate for most applications (40 dB and higher is more typical). Multiple FETs are chained together to achieve the required gain level. Figure 4.11 shows the layout of a generic two-stage amplifier.

In Figure 4.11, each stage has an independent gate and drain voltage. Often, each stage shares the same bias. This reduces the number of power supplies needed.

The steps in designing a multistage amplifier are listed as follows:

- Calculate the periphery needed on the output (final) stage to meet the power requirement.
- Determine the optimal bias and port impedances of the final stage to meet the other performance requirements.
- Calculate the periphery needed on the input (driver) stage. If high efficiency is needed from the two-stage amplifier, the input stage should have a much smaller periphery to minimize the DC power consumed (rule of thumb: 8:1 power ratio between the final and driver stage). If

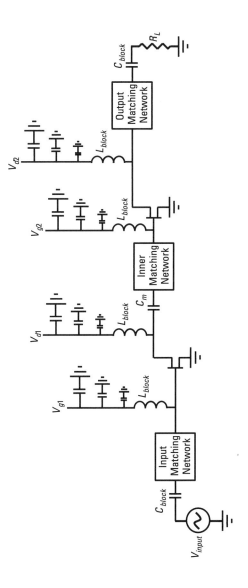

Figure 4.11 Two-stage amplifier with individual gate and drain bias.

high linearity is needed, the input stage should be closer in size to the final stage so it is linear when the final stage starts compressing (rule of thumb: 4:1 or 2:1 power ratio between the final and driver stage).

- Determine the optimal bias and port impedances of the driver stage to meet the other performance requirements.

- Design the inner-stage matching network that transforms the impedance needed at the input to the final stage to the impedance needed at the output of the driver stage. Be sure a series capacitor (C_m) is part of the matching network to provide a DC block between the input-stage drain and the output-stage gate voltage.

- In simulation, vary (or optimize) the inner-stage matching network to ensure mismatch is minimized and optimal impedances are presented.

It is not uncommon to lose several decibels in an inner-stage matching network. This is especially true if there is a large impedance difference between stages or if a broadband match is needed.

4.3 Broadband Amplifiers

The terms broadband, wideband, and wide instantaneous bandwidth are interchangeable and mean different things to different industries. In communications, broadband can mean 100s of megahertz. In electronic warfare, broadband can mean 50:1 or 100:1 bandwidth ratios relative to the operating frequency. Ultra-wideband (UWB) radar bandwidths tend to fall in between those extremes (typically 10:1 to 20:1 or less). By comparison, the vast majority of today's radars have an instantaneous bandwidth of ~10%.

For a long time, radars seemed to be *banded*, meaning that there were S-band radars, X-band radars, and Ku-band radars, among others. Today, there is a push to design, produce, qualify, and deploy one hardware design that fits as many systems or applications as possible. This requires broadband amplifier designs, which are possible but introduce the following trade-offs:

- For maximum gain and output power, amplifiers are conjugate-matched. In practice, this can only be realized narrowband so some performance will be sacrificed broadband.

- Broadband circuits will inherently have worse return loss over narrowband (explained by the Bode-Fano limit in Chapter 6).

- Rather than utilizing reactive matching, which is frequency-dependent, resistive matching is used. This reduces gain and increases NF.

• The natural amplifier gain slope is -6 dB/octave, so more octaves covered means that it is necessary to compensate for more gain slope.

There are four popular methods for achieving broadband performance: multisection matching networks, balanced amplifiers, push-pull amplifiers, and distributed amplifiers. Section 4.3.1–4.3.4 discuss each of these.

4.3.1 Multisection Matching

Any two impedances can be matched with two elements at a single frequency. Since inductors, capacitors, and transmission lines are frequency-dependent, the impedance they present are not maintained as bandwidth increases. Resistors can be used as broadband matching elements since they are not frequency-dependent (parasitic effects aside), but they are lossy.

Matching networks are broadband if they have low quality factors, which means that the impedance transformation remains close to the real line of the Smith chart ($R+j0\Omega$). In order to match two distant impedance, multiple elements are needed. To achieve the broadest bandwidth, elements should be chosen so the transformation *hops* along the real line. Figure 4.12 shows two 5Ω to 50Ω transformations centered at 10 GHz.

The impedance transformation can be made with a series 0.24-nH inductor and a shunt 0.95-pF capacitor. On the Smith chart, the lighter line shows the impedance trace, which is quite a distance from the real line. Alternatively, by using a sequence of series inductors and shunt capacitors and a transmission line, the same impedance match can be made. On the Smith chart, the darker line shows the impedance trace, which is much closer to the real line. Figure 4.13 shows the frequency response. The two-element matching circuit (dotted line) has a 10-dB return loss bandwidth of 2.4 GHz (24% bandwidth).

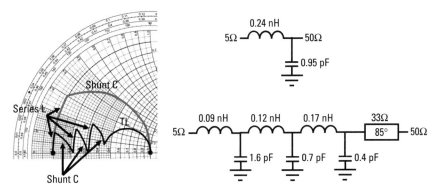

Figure 4.12 Two-element and seven-element matching circuit transformations from 5Ω to 50Ω to achieve broadband performance (centered at 10 GHz).

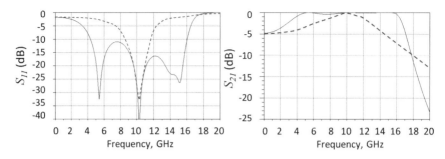

Figure 4.13 S_{11} and S_{21} of the circuits shown in Figure 4.12.

The seven-element matching circuit (solid line) has a bandwidth of 11.7 GHz (117% bandwidth), which is nearly five times the two-element bandwidth.

The previous example used a combination of lumped and distributed elements. The same could be accomplished with one or the other. At low frequency, lumped elements tend to be favored since parasitics are low and distribute elements are large. At high frequency, distributed elements are favored for the opposite reasons. For this example at X-band, either is applicable.

One additional benefit of broadband distributed matching circuits is their reproducibility in production. During fabrication, manufacturers can tightly control the line widths to achieve the designed line impedance. The unit-to-unit variation can be much better with distributed elements. Lumped element providers bin parts by value and can also tightly control the nominal value, but parasitics can vary considerably (discussed in Chapter 7). Figure 4.14 shows 5Ω to 50Ω distributed transformations centered at 10 GHz.

For the one-element case, the transmission line impedance is $\sqrt{5 \cdot 50} = 15.8\Omega$. For the two- and three-element cases, the impedance is chosen so that the impedance path remains as close to the real line as possible. Figure 4.15 shows the bandwidths for a one-, two-, three-, and four-element distributed matching network.

4.3.2   Balanced Amplifier

Balanced amplifiers are comprised of two identical amplifiers combined in parallel using two hybrid (90°) couplers (discussed in Chapter 6). Figure 4.16 shows a circuit schematic.

Balanced amplifiers can be used to ensure excellent return loss regardless of the reflection coefficient of the individual amplifiers. Figure 4.17 illustrates this concept. The incoming signal voltage is split equally ($V/2$) between the two amplifiers with 90° phase difference due to the couplers. The amplifiers will have reflected signals proportional to their reflection coefficients Γ_1 and Γ_2. The reflections to the input port will have 180° phase difference and cancel if

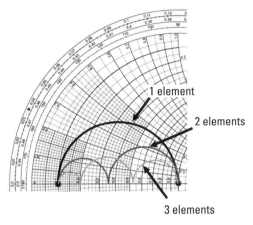

Figure 4.14 One-, two-, and three-element distributed matching transformations from 5Ω to 50Ω.

$\Gamma_1 = \Gamma_2$. Most of the reflected power is directed into the resistor at the isolation port, and only the portion resulting from the difference between Γ_1 and Γ_2 will appear at the input. The isolation resistor must be chosen carefully to ensure that it can handle the power level presented. Although in theory this principal works with any mismatch, to be effective in practice the return loss of each amplifier should be better than 5 dB.

Using this approach, designers can focus on optimizing the port impedances for power, efficiency, or linearity instead of impedance match. Since broadband impedances can be challenging to match, this approach mitigates that challenge.

The drawback to this approach is that it requires two amplifiers, two couplers, and two load resistors, so size and cost are higher. Broadband couplers are typically achieved using multiple stages, which increases size. The drawbacks are generally outweighed by advantages, so this is a popular design strategy.

4.3.3 Push-Pull Amplifier

Push-pull amplifiers are similar in appearance to balanced amplifiers, but they operate quite differently. Both circuits require two amplifiers, but in a push-pull circuit, they are combined using a balun (a structure that transforms a single port referenced to a common ground to a dual port referenced to each other—see Chapter 6). Figure 4.18 shows a push-pull amplifier.

Each amplifier receives half of the input signal, but separated by 180°. Baluns naturally have a lower port impedance on the splitting/combining side (12.5Ω or 25Ω generally), which means that the individual amplifiers only need to match to that lower impedance. This facilitates broadband matching, especially for very high-power amplifiers where impedance is low. Baluns also

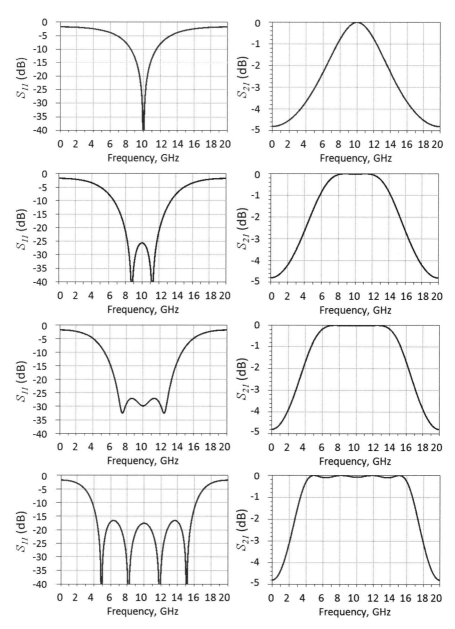

Figure 4.15 S_{11} and S_{21} of one-, two-, three-, and four-element distributed matching networks.

have the advantage of canceling second-harmonics. Chapter 6 explains this principle.

Unlike balanced amplifiers, baluns do not cancel reflected signals so individual amplifiers do need to be well matched. Isolation between amplifiers may

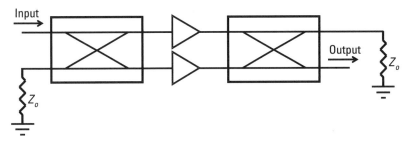

Figure 4.16 Circuit schematic of balanced amplifier.

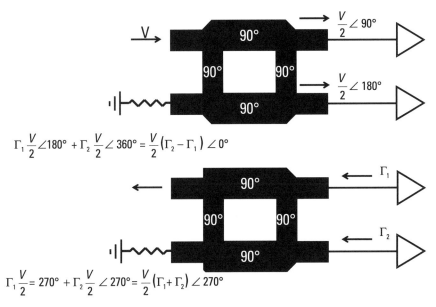

$$\Gamma_1 \frac{V}{2}\angle 180° + \Gamma_2 \frac{V}{2} \angle 360° = \frac{V}{2}\left(\Gamma_2 - \Gamma_1\right) \angle 0°$$

$$\Gamma_1 \frac{V}{2} = 270° + \Gamma_2 \frac{V}{2} \angle 270° = \frac{V}{2}\left(\Gamma_1 + \Gamma_2\right) \angle 270°$$

Figure 4.17 Explanation for cancellation of reflected signals in balanced amplifier.

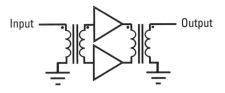

Figure 4.18 Circuit schematic of a push-pull amplifier

only be 6 dB with a balun, so a careful stability analysis must be performed. Some baluns utilize ferrites to achieve broad bandwidth, which can increase size and cost. As with the balanced amplifier, the drawbacks are generally outweighed by advantages so this is a popular design strategy.

4.3.4 Distributed Amplifiers

A distributed amplifier, illustrated in Figure 4.19, is a type of traveling-wave amplifier capable of achieving incredible bandwidths (much more so than the previous approaches discussed).

Multiple transistors are combined in parallel, although there is a cascading nature to the way the signals are combined. Transmission lines are placed between amplifiers to ensure in-phase combining. Load resistors (Z_d and Z_g) are added to absorb reflections.

Distributed amplifiers provide moderate match, gain, and noise figure. It's difficult to achieve very high power or efficiency, which can prohibit them from some applications.

4.4 Balancing Linearity and Efficiency

In the discussion on amplifier classes in Section 4.1, it was shown that amplifiers trade off efficiency for linearity. Class A provides excellent linearity and poor efficiency. Every class after that degrades in linearity but improves efficiency. Most applications require a balance between the two extremes. Class AB is one way of balancing linearity and efficiency. This section will discussion other options.

4.4.1 Explanation of Linearity

A circuit is no longer linear when increasing the input power by N dB does not also increase the output by N dB. That output energy must go somewhere, and some of it is generated as mixing products. This section will explain where these mixing products come from.

The voltage transfer function of an amplifier can be represented by the following Taylor series:

$$v_{out} = a_1 v_{in} + a_2 v_{in}^2 + a_3 v_{in}^3 + a_4 v_{in}^4 + \cdots \qquad (4.64)$$

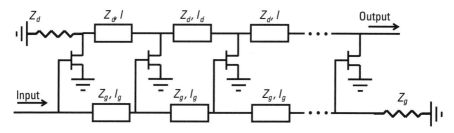

Figure 4.19 Circuit schematic of distributed amplifier.

where a_n is the coefficient associated with the nth term.

In the case where the input is a single frequency (neglecting phase):

$$v_{in}(t) = \cos(2\pi f t) \tag{4.65}$$

where f is the tone frequency (Hz) and t is time (s).

In the case where the input is multiple frequencies:

$$v_{in}(t) = \cos(2\pi f_1 t) + \cos(2\pi f_2 t) + \cdots \tag{4.66}$$

where f_n is the frequency of the nth tone.

Substituting $v_{in}(t)$ and solving for v_{out} gives:

$$
\begin{aligned}
v_{out} = {} & a_1 \left(\cos(2\pi f_1 t) + \cos(2\pi f_2 t)\right) \\
& + a_2 \left(\cos(2\pi f_1 t) + \cos(2\pi f_2 t)\right)^2 \\
& + a_3 \left(\cos(2\pi f_1 t) + \cos(2\pi f_2 t)\right)^3 \\
& + \cdots
\end{aligned}
\tag{4.67}
$$

The a_1 term produces f_1 and f_2 tones. Expanding the a_2 term provides:

$$
\begin{aligned}
& a_2 \left(\cos(2\pi f_1 t) + \cos(2\pi f_2 t)\right)^2 \\
& = a_2 \left(\cos^2(2\pi f_1 t) + 2\cos(2\pi f_1 t)\cos(2\pi f_2 t) + \cos^2(2\pi f_2 t)\right)
\end{aligned}
\tag{4.68}
$$

From trigonometric identities:

$$\cos^2\theta = \frac{1 + \cos 2\theta}{2} \tag{4.69}$$

$$\cos\theta\cos\phi = \frac{\cos(\theta - \phi) + \cos(\theta + \phi)}{2} \tag{4.70}$$

Therefore, we can expand to retrieve the individual frequency components:

$$a_2 \left(\begin{array}{l} \cos^2\left(2\pi f_1 t\right) + 2\cos\left(2\pi f_1 t\right)\cos\left(2\pi f_2 t\right) \\ + \cos^2\left(2\pi f_2 t\right) \end{array} \right)$$

$$= a_2 \left(\begin{array}{l} \dfrac{1 + \cos 2\left(2\pi f_1 t\right)}{2} \\[2mm] + \dfrac{\cos(2\pi f_1 t - 2\pi f_2 t) + \cos(2\pi f_1 t + 2\pi f_2 t)}{2} \\[2mm] + \dfrac{1 + \cos 2\left(2\pi f_2 t\right)}{2} \end{array} \right) \qquad (4.71)$$

$$= a_2 \left(\begin{array}{l} 1 + \dfrac{\cos 2\left(2\pi f_1 t\right)}{2} + \dfrac{1}{2}\cos(2\pi t(f_1 - f_2)) \\[2mm] + \dfrac{1}{2}\cos(2\pi t(f_1 + f_2)) + \dfrac{\cos 2\left(2\pi f_2 t\right)}{2} \end{array} \right)$$

The a_2 term generates $2f_1$, $f_1 - f_1$, $f_1 + f_2$, and $2f_2$ mixing products. Performing a similar analysis on the third term would generate mixing products at $3f_1$, $3f_2$, $2f_1 + f_2$, $f_1 + 2f_2$, $2f_1 - f_2$, and $-f_1 + 2f_2$.

Higher-order mixing products occur at frequencies, $mf_1 + nf_2$ where m and n may be positive or negative numbers. The order of the mixing product is defined as $|m| + |n|$. Figure 4.20 shows all of the mixing products for two tones, f_1 and f_2.

Practical Note

Another term for a mixing product is a "spurious signal." This is not to be confused with the term "harmonic," which is an integral multiple of a fundamental frequency (i.e., $2f_1$, $2f_2$, $3f_1$, and $3f_2$).

Practical Note

Although neighboring tones are shown in the Figure 4.20 as having equal amplitude, this is not necessarily the case (so $2f_1 - f_2$ doesn't have to equal $-f_1 + 2f_2$). For this reasons, mixing products are usually specified as "high-side" ($-f_1 + 2f_2$) or "low-side" ($2f_1 - f_2$).

Table 4.2 shows an example two-tone scenario at X-band with 1-MHz tone spacing. In a narrowband situation, all mixing products can be filtered except for the third-order tones at $2f_1 - f_2$ and $2f_2 - f_1$. They are extremely close in frequency to the two fundamental tones. For this reason, linearization efforts generally focus on IM3.

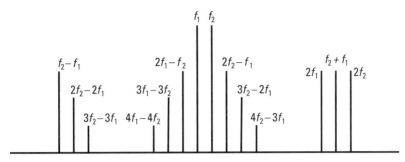

Figure 4.20 Mixing products for two input tones, f_1 and f_2.

Table 4.2
Mixing Products Generated from Two Input Tones
at 10 and 10.001 GHz

Mixing Products (gigahertz)					
First		**Second**		**Third**	
f_1	10	$2f_1$	20	$3f_1$	30
f_2	10.001	$2f_2$	20.002	$3f_2$	30.003
		f_2-f_1	0.001	$2f_1+f_2$	30.001
		f_2+f_1	20.001	$2f_2+f_1$	30.002
				$2f_1-f_2$	9.999
				$2f_2-f_1$	10.002

The levels shown in Figure 4.20 are typically seen when a nonlinear device (such as an amplifier) is several decibels into compression. At small-signal, when the device is linear, mixing products are not present and only the fundamental tones (f_1 and f_2 in this case) are present. As the input power level increases, the fundamental tones increase linearly (1-dB output: 1-dB input). However, second-order products increase by 2 dB for every 1 dB additional input power. Third-order products increase by 3 dB for every 1 dB additional input power (and so on). Mixing product levels are negligible at small-signal and will appear once the fundamental tones stop behaving linearly. If input power is increased high enough, mixing-product amplitude will meet or exceed the fundamental level. This is shown in Figure 4.21.

The circuit has a linear region with a 1-dB/dB slope. At some power level, the circuit will saturate (P_{sat}), and additional input power will generate no additional output power. The second-order mixing products (IM2) are summed and plotted with a 2-dB/dB slope. Intermodulation products are calculated by

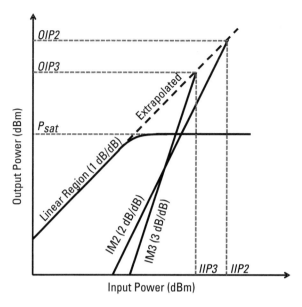

Figure 4.21　Graphical depiction of the various linearity terms.

measuring the amplitude and dividing the fundamental level with the mixing product level. The equation for calculating IM3 is

$$IM3(dBc) = \frac{P_{2f_2-f_1}}{P_{f_2}} = \frac{P_{2f_1-f_2}}{P_{f_1}} \tag{4.72}$$

where P_x is the power level at tone x. Units are in *decibels with respect to carrier.*

The IM2 line intersects a line extrapolating the power of the fundamental tone at a point called the *second-order intercept point* (IP2). That point corresponds to an input power IIP2 and an output power OIP2. The same process is done for the third-order products (IM3) and their related *third-order intercept point* (IP3).

Although intercept points cannot be measured, they are sometimes used as a way of specifying linearity. Often, OIP3 is ~10 dB above P_{sat} so that can be used as a rule of thumb.

Practical Note

In reality, mixing products compress and saturate just like the fundamental frequency. This can add ambiguity when specifying an intercept point since it is a calculated number. Since IMX are measured values, they are a safer way to specify linearity.

4.4.2 Doherty

Doherty circuits are advantageous when the average power is much less than the peak power (by 6–12 dB). When an amplifier operates backed off from saturation, the output is linear, but efficiency is low (efficiency usually peaks ~3–5 dB in compression). Doherty combines two amplifiers operating with different classes to achieve high efficiency at back-off and at saturation. Figure 4.22 shows the Doherty circuit.

The carrier amplifier is usually biased Class AB so it strikes a balance of linearity and efficiency. The peaking amplifier is biased Class C, so it is essentially OFF at back-off. The operation is shown in Figure 4.23.

The two amplifiers are separated by 90°, which can be implemented by a simple transmission line or a hybrid coupler (as shown in Figure 4.22). The advantage of a coupler is the cancelation of reflected signals at the input. At back-off, only the carrier amplifier is operational so its individual performance defines the Doherty output. The peaking amplifier is designed so it turns on at the same power level where the carrier amplifier starts to compress. This extends the linear operating region of the amplifier, so it is linear over a wide dynamic range. Concurrently, both the carrier and peaking amplifiers are running efficiently so the Doherty amplifier is efficient over the same dynamic range (the carrier defines the efficiency below saturation and the peaking defines the efficiency in saturation). On the output of the amplifiers, a quarter-wave transformer combines the signals in-phase. Since the effective impedance at that combination point is $Z_o/2$, another quarter-wave transformer is used to achieve Z_o.

This circuit approach has been widely used throughout the communications industry for decades. Many variations on this circuit have been made over the years to improve on performance, including the following:

- Unequal input power division between carrier and peaking;
- Incorporating baluns to cancel second harmonics;
- Using amplifiers with different output power;
- Combining more than two amplifiers.

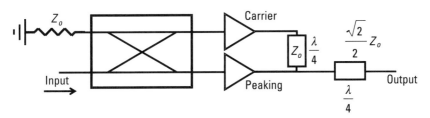

Figure 4.22 Circuit schematic of Doherty circuit.

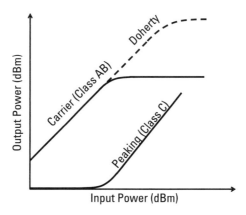

Figure 4.23 Output power versus input power for the carrier and peaking amplifiers within a Doherty amplifier (not drawn to scale to show alignment of power levels).

4.4.3 Other Linearization Techniques

The design approaches discussed to this point represent the industry standard, but there are some lesser known but equally beneficial options. These include the following:

1. *Transistor selection:* Transistors are made in foundries all over the world, and often they are chosen for convenience, price, availability, familiarity, and periphery, among other advantages. Linearity should also be a screening criteria during transistor selection. Device architecture and chemistry can inherently improve linearity by as much as 20 dB. A strong correlation has been found between doping profile and linearity, so some foundries have released specialized high-linearity processes. Recently, an emerging technology has made high-linearity GaN transistors commercially available.

Practical Note

When reviewing data sheets of transistors, be wary of "specmanship." Some transistor data sheets show linearity 10-dB backed-off from saturation. The linearity looks terrific, but they are operating in the linear region. Look for linearity plots versus output power instead of input power.

---

2. *Source and load matching:* Before designing, generate source and load-pull contours for linearity (i.e., IM3, IM2, and ACPR). This can be done in measurement or simulation. On one Smith chart, contours for power, efficiency, and linearity (not to mention noise, VSWR, and other parameters) can be plotted to directly compare the trade space.

3. *Analog predistortion* (APD): Predistortion is the process of modifying a signal before it enters the amplifier to correct for the expected nonlinearity. APD implements the modification using RF circuit elements, rather than digital processors. Figure 4.24 shows the predistortion process.

An amplifier has a set output power versus input power compression curve. A predistortion element is placed before the amplifier that has the complementary response. When the amplifier is linear, the linearizer has a fixed loss. When the amplifier starts to compress, the linearizer loss decreases at the same rate. Once the amplifier is saturated, the linearizer loss should ideally be zero. Effectively, this extends the linear region to a wider dynamic range. However, it does not increase the saturated output power level as is commonly shown in open literature. The watts per millimeter generated by an amplifier cannot increase due to the addition of a lossy element; therefore saturated output power cannot increase.

Analog predistorters are simple to integrate and can easily improve linearity by 10–20 dB (meaning that mixing-product amplitudes are 10–20 dB lower with a predistortion circuit added). The trade-off is added loss at back-off, which reduces gain and increases noise figure. They can be challenging to implement broadband since compression curve shape can change over frequency. If the linearizer loss and compression curves are not aligned, linearity can degrade instead of improve.

In production, APD can be challenging if there is a large variation from amplifier to amplifier. One way of mitigating this is by binning amplifiers and linearizers and matching curves.

4. *Digital predistortion* (DPD): DPD is similar to APD, except that the desired input to the amplifier is implemented by a synthesizer generat-

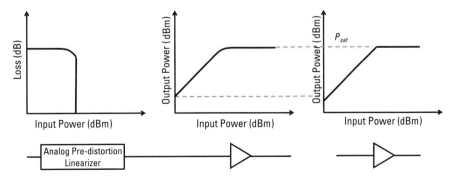

Figure 4.24   Predistortion process for a nonlinear amplifier.

ing the signal. Algorithms can be used to tune the waveform for each amplifier, which can improve linearity by 50–80 dB (same definition as before). Tuning in real time also makes DPD more effective across operating condition, broad bandwidth, and production variation.

Since radars have digital circuitry for performing complex mathematics, DPD is a more logical choice for linearization. This topic is explored further in Chapter 8.

4.5 Multiphysics Concerns

Cramming incredible amounts of power into a small area does you no good if the unit overheats and fails. Achieving never-before-seen efficiency is wasted if the material properties change when exposed to moisture. The best linearizer is deemed ineffective if the amplifier dies after 1,000 hours from a buildup of microcracks.

Today's radars are smaller, lighter, and higher-power, and they operate in the most extreme conditions ever. Designing for optimal electrical performance without considering mechanical concerns is a fool's errand and is a topic often overlooked in RF design books. This section discusses the mechanical concerns every radar component designer needs to know.

4.5.1 Thermal Considerations

Just as RF designers simulate circuits to evaluate the effects of loss, parasitics, and other imperfections, there is no substitute for a full thermal analysis. Multimaterial stack-ups spread heat in complicated ways. Tolerances in materials can be wide. Interfaces are never perfect. The effects of thermal conduction, convection, and radiation must be evaluated. In any system, there are multiple sources and sinks of thermal energy. The only way to truly capture this is in high-fidelity simulation and testing.

The thermal energy dissipated from an amplifier is defined by [1]:

$$P_{therm,diss} = \frac{T_j - T_A}{\theta_{JC} + \theta_{CS} + \theta_{SA}} \qquad (4.73)$$

where $P_{therm,diss}$ is the dissipated power (watts), T_j is junction temperature (°C), T_A is the ambient temperature (°C), θ_{JC} is the junction-to-case thermal resistance (°C/W), θ_{CS} is the case-to-heat sink thermal resistance (°C/W), and θ_{SA} is the heat sink-to-ambient thermal resistance (°C/W).

NOTE: T_j may also be called T_{jc} or T_c, which indicates that the device has a channel rather than a junction.

> **Practical Note**
>
> A good rule of thumb is that output power drops by 0.016 dB/°C per stage. So, an amplifier loses 0.5 dB for every ~30°C rise in temperature per stage.

Coefficient of Thermal Expansion

Most materials expand with increasing temperature and contract with decreasing temperature. The magnitude of that change, which is determined by a material's coefficient of thermal expansion (CTE), is represented by the following equation:

$$\varepsilon_T = \alpha \Delta T \tag{4.74}$$

where is the strain (change in size) caused by temperature (unitless), α is the CTE (1/°C), and ΔT is the change in temperature (°C). α is positive for expansion and negative for contraction.

The change in length due to change in temperature can be calculated by [4]:

$$\delta_T = \varepsilon_T L = \alpha \Delta T \cdot L \tag{4.75}$$

where δ_T is the change in length (m), and L is the original length (m).

Heat Capacity

Some objects, like a cast iron skillet, heat up very quickly when placed on a heat source. This is because they have a very low heat capacity, meaning that the application of very little heat results in a large temperature increase. This is represented mathematically as:

$$C = \frac{Q}{\Delta T} \tag{4.76}$$

where C is the heat capacity (calories per degree Celsius or joules per degree Celsius), Q is the heat applied (calories or joules), and ΔT is the change in temperature.

Heat capacity is proportional to mass (a larger skillet will have a smaller temperature rise than a small skillet for a fixed heat applied), which that means every object has its own value. This can make comparing multiple materials difficult. To avoid this, heat capacity is generally normalized to mass to make it a material-specific property, and this is known as the *specific heat*.

$$c = \frac{Q}{m\Delta T} \tag{4.77}$$

Specific heat is generally expressed as either cal/g $\cdot$ °C or J/Kg $\cdot$K.

Conduction

The rate of heat transfer for conduction is [5]:

$$H_{cond} = \frac{Q}{t} = kA\frac{\Delta T}{L} \tag{4.78}$$

where H_{cond} is the rate of heat transfer (W), k is the thermal conductivity (W/m $\cdot$K), A is the contact area (m²), L is the thickness (m), and ΔT is the change in temperature (°C).

The absolute thermal resistance due to conduction ($R_{th,cond}$) is defined as:

$$R_{th,cond} = \frac{L}{A \cdot k} \tag{4.79}$$

where L is the thickness (m) and A is the area of the object (m²).

Absolute thermal resistance is object-specific, and units are Kelvin per Watt (K/W). The generic thermal resistance of a material is calculated the same without dividing by the area. Units are meters squared Kelvin per Watt (m²$\cdot$K/W) or sometimes they are quoted in data sheets as feet squared Celsius hours per British thermal unit (ft²$\cdot$°C$\cdot$hr/Btu).

Practical Note

An often underestimated source of thermal resistance is due to imperfect contact between conducting surfaces. Even though the physical size may be small, surface roughness or contamination can have a significant effect. For example, the addition of a thin soft-metal foil (i.e., indium, lead, tin, or silver) or thermally conductive (i.e., silicon-based) paste between two surfaces can improve thermal conduction by four to 10 times.

Generally for microwave applications (especially power amplifiers), what matters most is the rise (or fall) in temperature as a result of a heat (or cooling) source. This can be calculated from:

$$\Delta T = \frac{H_{cond}R_{th,cond}}{A} \tag{4.80}$$

where ΔT is the change in temperature (°C), H_{cond} is the heat generated from the amplifier (W), $R_{th,cond}$ is the thermal resistance to the heat sink, and A is the contact area (m²).

For a multilayer, multimaterial, or composite material, the rise is temperature is calculated by summing the individual contributions to the thermal resistance.

$$\Delta T = \frac{H_{cond}\sum R_{th,cond}}{A} = \frac{H_{cond}\sum\frac{t}{k}}{A} \qquad (4.81)$$

where t is the thickness of the material (m) and k is the thermal conductivity Watts per Kelvin (W/m·K).

Convection

The heat flux for convection is:

$$Q = h\left(T_s - T_f\right) \qquad (4.82)$$

where Q has watts per square meter, h is the convection heat transfer coefficient Watts per meter squared Kelvin (W/m²·K), T_s is the surface temperature, and T_f is the fluid temperature. Table 4.3 lists typical ranges of h.

The large range of h shown in Table 4.3 is due to the complexity of moving fluids. Consider, for example, the following [4]:

- Objects "in the way" cause uneven flow.
- Fluids compress so there are areas of higher/lower pressure.
- Fluids experience inertia, which must be accounted for.
- Even on a smooth surface, friction slows the flow over the surface (that is, it has drag).
- Excessive disruptions in fluid flow causes turbulence.

Table 4.3
Range of Typical Convection Heat
Transfer Coefficients, h [5]

Convection Method	h (W/m²·K)
Free–gas	2–25
Free–liquid	50–1,000
Forced–gas	25–250
Forced–liquid	100–20,000

The rate of heat transfer for convection is [5]:

$$H_{conv} = QA = hA\left(T_s - T_f\right)$$ (4.83)

where H_{conv} has units (W).

The thermal resistance due to convection ($R_{th,conv}$) is defined as:

$$R_{th,conv} = \frac{1}{hA}$$ (4.84)

where h is the convection heat transfer coefficient (W/m²·K) and A is the area (m²). The units on $R_{th,conv}$ are also K/W.

Thermal Radiation

All objects above absolute zero (0K) radiate heat. Although the effects of thermal radiation are generally negligible for microwave components, they are included here for completeness. The power radiated due to its heat can be calculated by [5]:

$$H_{rad} = \sigma_{SB} \varepsilon_{rad} A\left(T_s^4 - T_{surroundings}^4\right)$$ (4.85)

where σ_{SB} is the Stefan-Boltzmann constant (5.6703×10^{-8} W/m²·K⁴), ε_{rad} is the emissivity (material property between 0 and 1), A is the surface area (square meters), T_s is the object surface temperature (K), and $T_{surroundings}$ is the temperature of the surrounding volume (K). In other text books, the subscripts "SB" and "rad" are not used, but since σ and ε have meaning in the microwave domain, they are included here for clarity.

Emissivity is a radiative property of the surface to quantify how efficient a surface radiates energy. The thermal resistance due to radiation ($R_{th,rad}$) is defined as:

$$R_{th,rad} = \frac{1}{\sigma_{SB} \varepsilon_{rad}\left(T_s + T_{surroundings}\right)\left(T_s^2 + T_{surroundings}^2\right)}$$ (4.86)

The effects of radiation are minimal. Figure 4.25 shows the radiated heat from a 6 mm × 6 mm GaN-on-SiC MMIC with the surrounding temperature set to ambient (25°C) as the surface temperature (T_s) is swept from 0K (−273.15°C) to 400K (126.85°C).

The dissipated heat is less than 32 mW. Given that a GaN MMIC that size is easily capable of generating more than 100W, this radiated power level is negligible (0.032%).

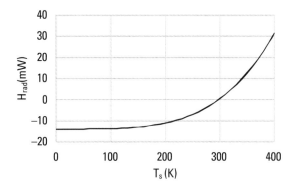

Figure 4.25 Radiated heat from a 6 mm X 6 mm GaN-on-SiC MMIC with the surrounding temperature set to ambient (25°C).

Cooling Techniques

The total rate of heat transfer from an object can be calculated by summing the conduction, convection, and radiation heat [5]:

$$
H = H_{cond} + H_{conv} + H_{rad} = \frac{A \Delta T}{R_{th,cond}}
$$
$$
+ hA\left(T_s - T_f\right) + \sigma_{SB}\varepsilon_{rad}A\left(T_s^4 - T_{surroundings}^4\right)
$$

(4.87)

In all cases, heat transfer can be maximized by increasing the contact area, but this conflicts with the objective of making circuits as small as possible. Choosing carrier materials to minimize the thermal resistance ($R_{th,cond}$) can improve heat transfer through conduction. Using forced air or liquid cooling can increase the convection heat transfer coefficient (h), although this is usually limited by the system.

Alternatively, fins can be added to aid in convection, and the *efficiency* is a measure of its effectiveness. The efficiency of a straight, rectangular fin can be calculated from [5]:

$$
\eta_{f,rect} = \frac{\tanh\left[\sqrt{\frac{2h}{kt}} \cdot \left(L + \frac{t}{2}\right)\right]}{\sqrt{\frac{2h}{kt}} \cdot \left(L + \frac{t}{2}\right)}
$$

(4.88)

where t is the fin thickness (m), L is the fin length (m), h is the convection heat transfer coefficient (W/m²·K), and k is the thermal conductivity (W/m·K) of the fin material. The range of $\eta_{f,rect}$ is between 0 and 1, where 1 is ideal.

The thermal resistance with the fin structure can be calculated from [5]:

$$R_{th,f,rect} = \frac{1}{2wh\eta_{f,rect}\left(L+\dfrac{t}{2}\right)}$$

(4.89)

where w is the fin width (m), t is the fin thickness (m), and L is the fin length (m). $R_{th,f,rect}$ is best (minimized) when $\eta_f = 1$.

Since the surface is partially covered in fins and partially unfinned, the total convection heat transfer is the sum of both surfaces:

$$H_{conv,f} = h\left(T_s - T_f\right)\left[2wN\eta_{f,rect}\left(L+\frac{t}{2}\right)+\left(A_t - 2wN\left(L+\frac{t}{2}\right)\right)\right]$$

(4.90)

where N is the number of fins and A_t is the total object surface area (i.e., overall length width if the object is rectangular) (square meters).

In comparison, the efficiency of a straight pin or cylindrical fin is determined by [5]:

$$\eta_{f,pin} = \frac{\tanh\left[\sqrt{\dfrac{4h}{kD}}\cdot\left(L+\dfrac{D}{4}\right)\right]}{\sqrt{\dfrac{4h}{kD}}\cdot\left(L+\dfrac{D}{4}\right)}$$

(4.91)

where D is the pin diameter (m) and L is the pin length (m).

The thermal resistance can be calculated from [5]:

$$R_{th,f,pin} = \frac{1}{\pi Dh\eta_{f,pin}\left(L+\dfrac{D}{4}\right)}$$

(4.92)

where D is the pin diameter (m) and L is the pin length (m).

The total convection heat transfer due to the finned and unfinned surface is:

$$H_{conv,f} = h\left(T_s - T_f\right)\left[\pi DN\eta_{f,pin}\left(L + \frac{D}{4}\right) + \left(A_t - N\pi D\left(L + \frac{D}{4}\right)\right)\right] \quad (4.93)$$

Figure 4.26 shows a plot comparing the efficiency of 1-cm-long aluminum fins with forced-air cooling. If either is more than a few millimeters thick, the difference is almost negligible. In both cases, forced air cooling with a fin is very effective at removing heat.

Thermal Spreaders

Sometimes designers do not have the luxury of a heat sink in close proximity to a heat source. In these cases, it's important to spread the heat as much as possible to reduce peak temperature. Fortunately, there are a number of commercial and military-grade materials that can be used for this. Some examples are listed as follows:

- Thermally conductive pastes, gels, and greases;
- Silicone and nonsilicone thermal sheets to serve as interface layers;
- Heat pipes;
- High thermal-conductivity ceramics;
- High thermal-conductivity semiconductor (i.e., aluminum nitride);

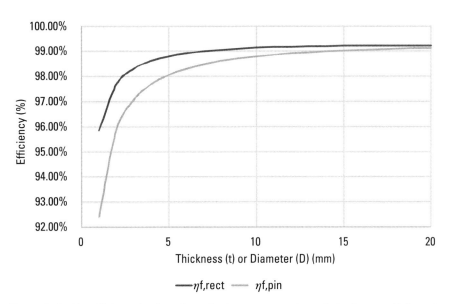

Figure 4.26 Fin efficiency for 1-cm-long aluminum rectangular and pin shapes with forced-air cooling.

• Chemical vapor deposition (CVD) diamond.

Generally these approaches are implemented at the module level, so they are discussed in Chapter 8.

4.5.2 Mechanical Considerations

The real world is not static. Most if not all commercial and military radars go through some degree of "shake, rattle, and roll" testing. This is to verify that the design can handle the effects of the operating environment. These conditions must be considered in the microwave design. There is no substitute for a complete mechanical analysis, but this section will review the key terms and equations you need to know [6]. Again, high-fidelity simulations can greatly aid designers and reduce the overall time needed to complete a successful design.

> **Practical Note**
>
> In an informal survey conducted at the 2014 International Microwave Symposium, a significant percentage of microwave designers admitted to spending as much time accounting for thermal and mechanical requirements as electrical.

Stress

Stress is the intensity of a force applied to an object. The magnitude of the stress is defined as:

$$\sigma = \frac{P}{A} \tag{4.94}$$

where σ is the stress [pounds per square inch (psi), Newtons per square meter (N/m^2), or Pascals (Pa)], P is the force or pressure (pounds or Newtons), and A is the area over which the force is applied (in^2 or m^2).

If the force is in a direction that stretches the object, it is called *tensile stress* and the sign on σ is positive. If the direction compression the object, it is called *compressive stress* and the sign on σ is negative.

Strain

Some objects are brittle in nature and do not deform under stress. Many ceramics, for example, hold their shape until enough stress is applied to cause cracking. Other materials, like metals, are ductile in nature. Their shape will deform under stress. When an object elongates under tensile stress or shortens under compressive stress, this change in length is call strain. It is defined as [4]:

$$\varepsilon = \frac{\delta}{L} \tag{4.95}$$

where ε is the strain (unitless), Δ is the change in length (m), and L is the original length (m). Often, strain is simply expressed as a percentage.

If the object is stretched, it is called *tensile strain* and the sign on ε is positive. If the object is shortened, it is called *compressive strain* and the sign on ε is negative.

Most materials do not change in just one direction when stretched or compressed. If you stretch a bar of metal, for example, the length will elongate, but the width will also decrease (you get a longer, thinner wire). That ratio of length elongation to width narrowing is known as *Poisson's ratio:*

$$v = -\frac{\varepsilon_{lateral}}{\varepsilon_{axial}} \tag{4.96}$$

where v is Poisson's ratio (unitless), $\varepsilon_{lateral}$ is the strain in the direction normal to the applied stress (unitless), and ε_{axial} is the strain in the direction parallel to the applied stress (unitless).

Generally, either $\varepsilon_{lateral}$ or ε_{axial} is negative, but not both (that would imply growth in all directions when stretched, for example). Therefore, to make Poisson's ratio positive, the convention is to add a negative sign to the equation.

A material can also be expressed by its modulus of elasticity (or *Young's modulus*) defined by:

$$E = \frac{\sigma}{\varepsilon} \tag{4.97}$$

Since strain is unitless, E is expressed as the same units as stress. This equation is commonly known as *Hooke's law.* Using this value, we can calculate the stress caused by a change in temperature by:

$$\sigma = E\alpha\Delta T \tag{4.98}$$

where α is the coefficient of thermal expansion (1/°C).

Fatigue

Over its lifetime, a radar component is exposed to a highly dynamic environment. Radar systems may be transported and deployed many times. Desert deployment can mean very hot days and very cool nights. Radars will cycle between moments of activity and rest. Even during constant operation, conditions within the radar may change (i.e., adaptive output power) to meet performance needs. The effect of these individual stresses may be small (almost negligible), but when factored over the lifetime of the system (10–30 years, or longer) they can be significant. The wearing down or deterioration of components due to repetitious stress is called *fatigue*. It's not uncommon for radar components to experience millions or billions of repeated stress cycles over their lifetime. Fatigue can occur in metal, polymers, and ceramics, although ceramics are the least susceptible to fatigue [7].

Practical Note

Taking a paper clip and bending it back and forth until it weakens and eventually breaks is an example of failure due to fatigue.

Failure due to fatigue happens when the stress is so high that the mechanical structure loses integrity and a microcrack is formed. This reduces the area over which the stress is distributed, which concentrates (increases) the stress. If the stress condition continues, the microcrack will spread until the area intact is no longer sufficient to bear the stress. Generally with microwave components, materials are small and thin so once a microcrack is formed, catastrophic failure happens quickly.

Figure 4.27 shows a representative *endurance curve* (or S-N curve). It indicates two important relationships between stress and endurance:

- There is an upper level of stress under which the object cannot endure even one cycle;
- There is a lower level of stress under which the object could (theoretically) endure an infinite number of cycles (called the *endurance limit*) [7].

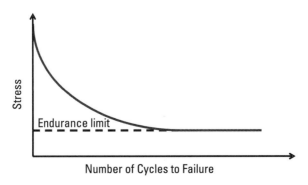

Figure 4.27   Example endurance curve.

> **Practical Note**
>
> Not all materials have a clearly defined endurance limit as shown in Figure 4.27. Some materials exhibit S-N curves that continue to decrease instead of flatten. In these cases, it is generally accepted to set the endurance limit at the stress level corresponding to 10^7 cycles.

The properties of fatigue are difficult to quantify because they vary with the following factors [6]:

- Properties of the material(s) involved;
- Geometry of the components;
- Speed of stress changes;
- Magnitude of stress.

As a rule of thumb, the endurance limit is generally between 25 and 50% of the *ultimate tensile strength* (σ_{uts}) of the material (a value provided by the material manufacturer).

Since fatigue always starts at the surface, components can be designed for greater endurance by optimizing the surface condition. Materials that are naturally rough (i.e., machined metal parts) are naturally prone to instigate microcracks. Polishing the surface can reduce the tendency to crack. Adding a surface treatment to prevent oxidation or corrosion can help resist fatigue. Tool or grinding marks should run parallel to the primary loading direction, if possible.

4.6 Local Oscillators (LOs)

Much of this chapter deals with how to prevent unstable or oscillating amplifiers. Otherwise, signals will be generated other than what is desired. However, there is a need for LOs, active components that intentionally oscillate to generate an RF signal. LOs are rated based on how constant the frequency remains over time and temperature (called *stability*), level of phase noise (fluctuations in the phase of a waveform), and linearity (no tones generated other than the intended one).

There are several types of LOs, including the following:

- Crystal oscillators: Components that use the mechanical resonance of a vibrating crystal (usually quartz) to create a very precise frequency. These are popular due to their good stability and low cost. However, crystal oscillators can only operate at one frequency.

- Voltage-controlled oscillators (VCOs): Tunable components that are capable of changing the oscillating frequency using tunable analog elements, such as varactors. At high frequency, tunable analog elements can have excessive parasitic effects that can limit performance.

- Direct digital synthesizers (DDS): Tunable components that are capable of changing the oscillating frequency using digital circuits. These components have the disadvantage of being relatively noisy and offer less tunability than analog circuits. However, a single integrated circuit can provide DSS, so size and cost are minimal.

To improve stability in tunable circuits, phase-locked loops (PLLs) can be incorporated. A PLL is a feedback mechanism that compares the input phase (and frequency) to the output phase (and frequency). It is possible to compensate for any detected discrepancy to maintain stable performance [8].

Practical Note

Phase noise and jitter are essentially the same thing. The former term is generally used in the RF domain, and the latter is used in the digital domain.

Chapter 6 shows how an LO can be combined with a mixer to perform up- and down-frequency conversion. LO design is outside the scope of this book, but interested readers are encouraged to review [9].

4.7   Tubes, Solid-State, and Where They Overlap

Over the past decade, solid-state sources have become the preferred method of generating RF power in radar transmitters, both in new designs and in retrofitting legacy radars. A large fraction of radars operating today still rely on tubes for the final power stages. However, tubes suffer from wear and tear issues, and they are being replaced with solid-state power amplifiers whenever possible. In the class of electronically scanned array antennas, solid-state transmit/receive modules now power the largest and most powerful ground-based radars (e.g., BMEWS and PAVE PAWS systems) and virtually all the latest multipurpose radars for fighter and attack aircraft.

The introduction of GaN and its 5–15 W/mm power density marked the first time solid-state amplifiers could go head-to-head with tubes. GaN-on-diamond, GaN-on-graphene, and other technologies are on the horizon and offer the potential of three to five times those levels (hence why nearly half of the power amplifier chapter was dedicated to thermal and mechanical considerations). Through advancements in radar technology, AESAs achieve high powers by combining many solid-state amplifiers (sometimes in the thousands).

The need for megawatt power for a single amplifier is not needed. There could come a day when solid-state capabilities equal or overcome tubes.

Nevertheless, tube manufacturers are not resting on their laurels. Tubes are getting smaller, more reliable, and more affordable. They still require electron generation, which typically includes a warm-up delay. Power supply voltages still range above 600V, which can be an issue—but these voltages are decreasing. Reliability is still measured in years instead of hundreds of years (like solid-state), but this is also improving.

Some of the best circuits utilize solid-state amplifiers to drive tubes. Although the focus of this book is on solid-state, at least for the foreseeable future, tubes will continue to have a place in radar systems.

Exercises

1. Draw the class A load line for the IV plane generated from Exercise 3.6. What is the maximum output power of that transistor?

2. Design a suitable DC block and RF choke for a 5-GHz amplifier with a 50-Ω system impedance.

3. A circuit has $S_{11} = 0.1$, $S_{22} = 0.08$, $S_{21} = 4$, and $S_{12} = 0.01$. Is it unconditionally stable?

4. Design a gain-flattening circuit capable of compensating for a -6-dB/octave (meaning that every time the frequency doubles, the gain drops by 6 dB) from 4 to 16 GHz.

5. Design a broadband impedance-matching network capable of matching 8+j20Ω to 50Ω. Repeat for 8–j20Ω. Center both networks at 10 GHz.

6. The output power of a GaN amplifier is 75W. What is the approximate OIP3?

7. An ideal 100-W two-stage class A amplifier in a 25°C lab has a thermal resistance of 0.65°C/W. How much will the junction temperature change if a heat sink is added that drops the thermal resistance to 0.40°C/W?

8. How much additional gain could be expected with this temperature change?

9. If the carrier beneath the amplifier is 5 mm × 5 mm × 0.2 mm aluminum, what will the size be as a result of the temperature change?

10. A radar submodule is 35°C above the temperature necessary for reliable operation. The module is 4 in × 8 in × 1 in with a 250-mil-thick aluminum base. What techniques could be used to reduce the op-

erating temperature without modifying the electronics? Choose one technique and design a modification to the submodule that meets the temperature specification. Calculate the expected drop in temperature.

11. An amplifier being designed does not have adequate gain. What can be done to the circuit to increase the maximum stable gain?

12. Use Keysight ADS to plot gain (or power), efficiency, VSWR, and stability circles for a transistor with default values. Alternatively, perform the same analysis using a transistor model from a commercial foundry.

References

[1] Chang, K., I. Bahl, and V. Nair, *RF and Microwave Circuit and Component Design for Wireless Systems,* New York, NY: John Wiley & Sons, 2002.

[2] Edwards, M. L., and J. H. Sinsky, "A New Criterion for Linear 2-Port Stability Using a Single Geometrically Derived Parameter," *IEEE Transactions on Microwave Theory and Techniques,* Vol. 40, No. 12, Dec. 1992, pp. 2303–2311.

[3] Collin, R., *Foundations for Microwave Engineering,* New York, NY: IEEE Press, 2001.

[4] Senturia, S., *Microsystem Design,* Norwell, MA: Kluwer Academic Publishers, 1940.

[5] Incropera, F., and D. DeWitt, *Fundamentals of Heat and Mass Transfer,* New York, NY: John Wiley & Sons, Inc., 2002.

[6] Gere, J., *Mechanics of Materials,* Pacific Grove, CA: Brooks/Cole, 2001.

[7] Schaffer, J., et al., *The Science and Design of Engineering Materials,* New York, NY: WCB McGraw-Hill, 1999.

[8] Kroupa, V., *Phase Lock Loops and Frequency Synthesis,* Chichester, England: John Wiley & Sons Ltd, 2003.

[9] Odyniec, M., *RF and Microwave Oscillator Design,* Norwood, MA: Artech House, 2002.

Selected Bibliography

Kenington, P., *High-Linearity RF Amplifier Design,* Norwood, MA: Artech House, 2000.

Maas, S., *Practical Microwave Circuits,* Norwood, MA: Artech House, 2014.

Vuolevi, J., and T. Rahkonen, *Distortion in RF Power Amplifiers,* Norwood, MA: Artech House, 2003.

5

LNAs

If the power amplifier is the cornerstone of the transmit radar, the LNA is the equivalent for the receive radar. Before reviewing LNA design strategies, it is important to understand what noise is and where it comes from. Noise is an electrical signal with random behavior. All electronic circuits inherently generate noise. Noise can be measured, and even though the amplitude is random, the average magnitude can be predicted by probability. Figure 5.1 shows what a real sine wave might look like with and without noise.

Sources of noise in a radar can be external to a system (i.e., received from *sky noise,* some of which—amazingly—is a remnant of the Big Bang!) or internal (i.e., created by the components within the system). Noise from one component can couple to another component. This is known as *interference* or *crosstalk* (discussed in Chapter 8).

This chapter introduces the types of noise present in radar systems and demonstrates how to calculate component performance with noise. In addition, it discusses the process for designing an LNA, including information on self-biasing, equalization, designing for high dynamic range, and cryogenic operation.

5.1   Explanation of Noise

There are several sources of noise related to radar components, including the following [1]:

- Thermal (also known as Johnson or Nyquist) noise: Generated by thermal energy from all resistive elements as free electrons move randomly at temperatures above absolute zero (0K).

- Shot noise (also known as Poisson noise): The result of random fluctuations of charge carriers across a transistor p-n junction, which is directly proportional to DC bias current.
- Flicker, 1/f (read "one over f"), contact, excess, or pink noise: Triggered when imperfect contact between two conducting materials causes conductivity to fluctuate when DC current is applied.
- Quantum noise: Caused by charge carriers and photons. The magnitude is generally negligible compared to the other noise sources in most devices.

The Fourier transform (5.1) is an integral equation that converts a signal in the time domain (like the one shown in Figure 5.1) into the frequency domain. It is given by:

$$G(f) = \int_{-\infty}^{\infty} g(t)e^{-j2\pi ft}\, dt \tag{5.1}$$

where $G(f)$ is the frequency-domain response, $g(t)$ is the time-domain response, j is the imaginary number $\sqrt{-1}$, f is the frequency of interest, and t is time. The frequency-domain response is determined by calculating (5.1) at every frequency point. The function of a spectrum analyzer is to perform this process fast enough to be displayed in real time.

Taking the Fourier transform of a signal determines the magnitude of the desired and undesired (including noise) frequencies. The frequency response after taking the Fourier transform of a single-tone signal is shown in Figure 5.2.

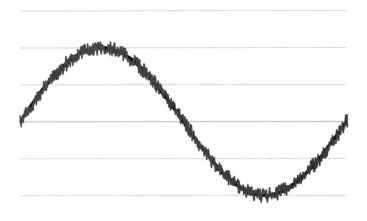

Figure 5.1   Sine wave with (gray trace) and without (black trace) noise.

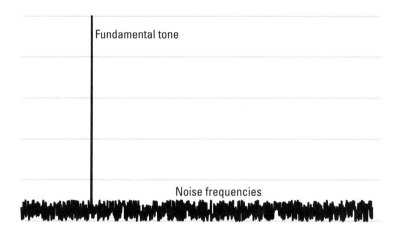

Figure 5.2 Frequency domain equivalent of Figure 5.1.

Figure 5.2 shows noise occurring at nearly every frequency. Integrating the noise level across all frequencies would give an infinite response. Therefore, the only way to quantify noise is to specify a bandwidth of interest. The amount of noise occurring in a given bandwidth is known as the *noise power spectral density*, or more commonly, *noise power density* (NPD). The units of NPD are typically watts per hertz (W/Hz) or watt-seconds (W-s), and the symbol is N_0. The wider the bandwidth, the more noise that will fall within that bandwidth.

5.1.1 Thermal Noise

The history of thermal noise (and noise in general) is presented in great detail by Leon Cohen in [2]. Thermal noise has a uniform or flat power distribution across frequencies giving it the distinction of being called *white noise* (an analog to white light, which encompasses all optical frequencies).

Thermal noise is the dominant source of noise in most radar systems operating above high-frequency (HF), where sky noise can actually be dominant. Thermal noise voltage can be calculated using Planck's black body radiation law (rms value):

$$V_n = \sqrt{\frac{4hfBR}{e^{hf/kT} - 1}} \tag{5.2}$$

where h is Planck's constant (6.626×10^{-34} J-sec), k is Boltzmann's constant (1.381×10^{-23} J/K), T is temperature (K), B is system bandwidth (Hz), f is the center frequency (Hz), and R is the noise resistance (Ω).

This equation can be simplified using Taylor series expansion for exponents:

$$e^x = \sum_{n=0}^{\infty} \frac{x^n}{n!} = 1 + x + \frac{x^2}{2!} + \frac{x^3}{3!} + \frac{x^4}{4!} + \cdots \text{ for all } x \qquad (5.3)$$

The exponent in the denominator of (5.2) can be simplified to:

$$e^{hf/kT} \approx 1 + \frac{hf}{kT} + \frac{(hf/kT)^2}{2!} + \frac{(hf/kT)^3}{3!} + \frac{(hf/kT)^4}{4!} + \cdots \qquad (5.4)$$

The ratio of h/k is approximately 4.74×10^{-11}. Since RF radar has frequencies on the order of GHz (10^9), the first term in the Taylor series is significant. In higher-order terms, that ratio trumps the frequency so as an approximation, they can be ignored. The exponent of (5.2) then simplifies to:

$$e^{hf/kT} \approx 1 + \frac{hf}{kT} \qquad (5.5)$$

Inserting this approximation into (5.2) gives:

$$V_n = \sqrt{\frac{4hfBR}{e^{hf/kT} - 1}} \approx \sqrt{\frac{4hfBR}{\left(1 + \dfrac{hf}{kT}\right) - 1}} = \sqrt{4kTBR} \qquad (5.6)$$

This is known as the *Rayleigh-Jeans approximation*. Notice that thermal noise voltage is independent of frequency as stated earlier, and it applies to the open-circuit voltage of resistance R at temperature T. Figure 5.3 shows the noise voltage for three bandwidths at 290K versus resistance.

Noise sources that may or may not be derived from temperature but that are not a function of frequency can be considered white noise and handled as such. When this is the case, noise temperature is referred to as the *equivalent noise temperature*, T_e [1], to differentiate from a pure thermal noise source.

From the Rayleigh-Jeans approximation, it was shown that noise is proportional to bandwidth and temperature; reducing one or both will reduce the noise level.

Noise power can be calculated from [3]:

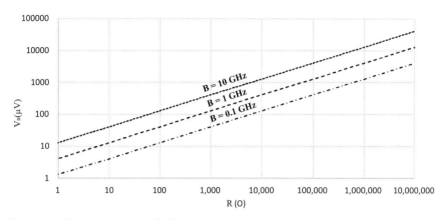

Figure 5.3 Noise voltage from (5.6) as resistance varies.

$$P_n = kTB \tag{5.7}$$

where P_n is the noise power (W), k is Boltzmann's constant (1.381×10^{-23} J/K), T is temperature (K), and B is system bandwidth (Hz).

From (5.7), noise power density can be calculated by:

$$N_o = \frac{P_n}{B} = kT \tag{5.8}$$

Practical Note

Although it's been stated that white noise occupies all frequencies, it actually has a cut-off frequency in the infrared range (300 GHz–430 THz). For all intents and purposes for radar, assume that white noise occupies all frequencies.

5.1.2 Shot Noise

Shot noise occurs in transistor devices with a junction (as opposed to a conducting channel, discussed in Chapter 3). This includes p-n junction diodes, bipolar transistors, and MOSFETs. It is caused by random diffusion of holes and electrons through a p-n junction and by the random generation and recombination of hole-electron pairs [4]. When this happens, random fluctuations of current occur. Shot noise current can be calculated by the Schottky formula [5]:

$$I_{shot} = \sqrt{2qI_{DC}B} \tag{5.9}$$

where q is the charge of an electron (1.6×10^{-19} C), I_{DC} is the average DC current (amperes), and B is the noise bandwidth (hertz).

Since shot noise is independent of frequency, it is also classified as white noise. For practical purposes, unless current or bandwidth are extremely high, shot noise tends to be negligible. For example, the shot noise generated from 1-A and 1-GHz bandwidth, is only $17.9\ \mu A$ (0.00179% of the 1-A source).

5.1.3   Flicker Noise

When DC current flows through two conductive materials, imperfections at the interface can generate noise current. This is called *flicker noise,* and it can be calculated by [4]:

$$I_{flicker} = \sqrt{\frac{K_f I^m B}{f^n}} \tag{5.10}$$

where K_f is the flicker noise coefficient (A), I is the DC current (A), m is the flicker noise exponent (usually $1 < m < 3$), B is the noise bandwidth (Hz), f is the frequency (Hz), and n is the frequency exponent (usually $n \approx 1$). K_f can be calculated by [4]:

$$K_f = \frac{16}{3} k T f_f \sqrt{\frac{I_{DSS}}{V_{po}^2 I_d}} \tag{5.11}$$

where k is Boltzmann's constant (1.381×10^{-23} J/K), T is temperature (K), f_f is the flicker noise corner frequency (Hz), I_{DSS} is the saturated current level where $V_g = 0$ (discussed in Section 3.3.2), V_{po} is the pinch-off voltage (V), and I_d is the drain current (A). The flicker noise corner frequency is defined as the frequency where the noise spectral density is 3 dB higher than the voltage at high frequency.

Notice that flicker noise is only generated when DC current flows. Since the noise level varies inversely proportional with frequency, it is called $1/f$ (read "one over f") *noise.* Since the noise level is higher at low frequency, it is also called *pink noise* because pink light has a power density that increases at lower frequency.

Practical Note

Above 100 MHz or so, flicker noise is insignificant compared to thermal and shot noise. For radar applications, it is easy to ignore this effect. However, low-frequency oscillations or mixing products (discussed in Chapter 8) can fall below 100 MHz and generate flicker noise.

5.1.4 Noise Terminology

A valid assumption is that the average voltage of a noise signal ($v_{n,avg}$) over time is zero. That is [6]:

$$v_{n,avg} = \lim_{x \to \infty} \frac{1}{x} \int_{-\frac{x}{2}}^{\frac{x}{2}} v_n(t)dt = 0 \qquad (5.12)$$

However, the average of the square of a noise signal is nonzero. This is known as the *mean-square noise,* and it is calculated from:

$$v_{n,avg}^2 = \lim_{x \to \infty} \frac{1}{x} \int_{-\frac{x}{2}}^{\frac{x}{2}} \left[v_n(t)\right]^2 dt \qquad (5.13)$$

Most commonly, noise is referenced to the square root of the mean-square noise, which is called the *root-mean-square* or *RMS noise* (denoted $v_{n,rms}$).

A term used by both system and component designers is *noise figure* (NF). The noise figure of a component is the change in the SNR if the component was noiseless (a noiseless component would have NF = 0 dB). This can be represented as:

$$NF(dB) = 10\log\left(\frac{S_i / N_i}{S_o / N_o}\right) \geq 1 \qquad (5.14)$$

where S is the signal level and N is the noise level. The subscript i denotes input and o denotes output. If the "10 log" is omitted from (5.14), the ratio is called the *noise factor* (F). Generally, designers think in terms of decibels so NF is a more popular term.

The SNR can be calculated from the signal and noise levels using:

$$SNR(dB) = 10\log\left(\frac{v_{s,avg}^2}{v_{n,avg}^2}\right) = 20\log\left(\frac{v_{s,rms}}{v_{n,rms}}\right) \qquad (5.15)$$

where $v_{s,avg}^2$ is the mean-square signal voltage, $v_{n,avg}^2$ is the mean-square noise voltage, $v_{s,rms}$ is the root-mean-square signal voltage, and $v_{n,rms}$ is the root-mean-square noise voltage.

From the NF, the equivalent noise temperature (or just *noise temperature*) can be calculated from:

$$T_e = T_o \left(10^{NF/10} - 1 \right)$$

(5.16)

with the NF in decibels and T_o is the standard noise temperature reference (defined as 290K).

Given the noise temperature, the NF can be calculated from:

$$NF(dB) = 10 \log \left(\frac{T_e}{T_o} + 1 \right)$$

(5.17)

where T_e is the noise temperature (K) and T_o is the standard noise temperature reference (K).

5.2 Transistor Noise Modeling

The noise performance of a transistor can be accurately modeled and simulated by adding four elements to the equivalent small-signal circuit model shown in Figure 3.6, listed as follows.

- Thermal noise voltage source generated by the gate resistance (V_{ng});
- Thermal noise voltage source generated by the source resistance (V_{ns});
- Shot noise current source generated by the gate (i_{ng});
- Shot noise current source generated by the drain (i_{nd}).

The enhanced circuit model with noise parameters is shown in Figure 5.4.

The equations for V_n (5.6) and I_{shot} (5.9) can be used as a starting point for the noise model. Unfortunately, the values (also called *model parameters*) for V_{ng}, V_{ns}, i_{ng}, and i_{nd} will need to be optimized to fit measured noise data. Material imperfections, parasitics, and other difficult-to-anticipate or difficult-to-quantify factors need to be included in the model parameters.

5.3 Design Strategies and Practices

To minimize noise, a low operating DC bias is needed. However, gain is reduced as DC bias is reduced due to the transconductance. Therefore, the principal trade-off made in LNA design is finding the optimal balance between noise and gain.

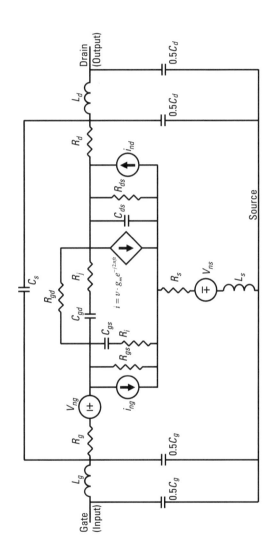

Figure 5.4 Equivalent circuit model with noise parameters added.

The signal received by a radar will be extremely low in amplitude, often just above the noise floor, which is the lowest measurable level that is distinguishable from background radiation (discussed in Section 5.4). To amplify the signal to a level that can be processed, an RF amplifier is needed that does not contribute significant additional noise. This is known as an *LNA*. Since the received signal may have a power level of only −60 to −100 dBm and a typical LNA stage has 10–20-dB gain, multiple LNAs must be cascaded to achieve the necessary power level. An LNA cascade is shown in Figure 5.5.

The noise factor of the cascaded chain can be determined using the Friis formula for noise:

$$F = F_1 + \left(F_2 - 1\right)\frac{M_2}{M_1 G_1} + \left(F_3 - 1\right)\frac{M_3}{M_1 M_2 G_1 G_2} + \cdots \qquad (5.18)$$

Since the overall gain increases with each additional stage, the denominator increases, which means that the noise contribution decreases.

Practical Note

The principal noise driver is the first stage, so it should be designed to have as low noise and high gain as possible. In order to overcome the noise of later stages, the first-stage gain must be at least 10 dB and have minimal NF—even if this means compromising the output return loss match.

5.3.1 Understanding Noise Circles

Similar to a power amplifier, the performance of an LNA is determined by how it is biased and how it is terminated. The optimal configuration can be determined by measuring a transistor with a noise parameter analyzer. The measurement system measures noise performance while varying the source impedance and generates four noise parameters at each bias:

- NF50: The NF at 50Ω;

- NF_{min}: The lowest NF possible;

- Γ_{opt} or Z_{opt}: The matching impedance where NFmin is achieved;

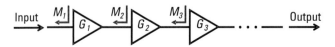

Figure 5.5 Cascade of LNAs with mismatch loss (M_N) and gain (G_N).

• Equivalent derivative noise resistance (R_e): The rate of change in noise performance as the impedance deviates from Γ_{opt}. Lower values of R_e indicate a wider range of impedances are possible with minimal degradation in NF.

By sweeping noise measurements over bias, a trade-off between NF and gain over a range of drain current can be made. This is shown in Figure 5.6.

Once the bias level where NF and gain are balanced is determined, noise circles can be drawn at that bias to determine the optimal terminating impedance. Plotting circles on a Smith chart provides designers with a visual tool to trade off noise performance with other parameters, such as stability, VSWR, or gain.

A conversion can be made between impedance and reflection coefficient using (5.19) where Z_o is the system impedance:

$$\Gamma_{opt} = \frac{Z_{opt} - Z_o}{Z_{opt} + Z_o} \tag{5.19}$$

For a given impedance of interest (Z_m), (5.19) can be used to calculate the reflection coefficient (Γ_m). From the four noise parameters, the noise factor at Γ_m can be calculated from [3]:

$$F = F_{min} + 4R_e \frac{\left|\Gamma_{opt} - \Gamma_m\right|^2}{\left|1 + \Gamma_m\right|^2 \left(1 - \left|\Gamma_{opt}\right|^2\right)} \tag{5.20}$$

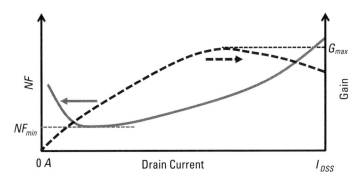

Figure 5.6 Example relationship between NF, gain, and drain current.

where F is the noise factor, F_{min} is the minimum noise factor, R_e is the equivalent derivative noise resistance, Γ_{opt} is the optimum noise reflection coefficient, and Γ_m is the reflection coefficient of interest. Recall that NF = 10 log (F).

To make calculations easier, a noise circle index parameter (N_{ci}) is defined as [3]:

$$N_{ci} = \frac{\left|\Gamma_{opt} - \Gamma_m\right|^2}{1 - \left|\Gamma_{opt}\right|^2} = \frac{\Gamma_{opt}\Gamma^*_{opt} - \Gamma_{opt}\Gamma^*_m - \Gamma_m\Gamma^*_{opt} + \Gamma_m\Gamma^*_m}{1 - \Gamma_{opt}\Gamma^*_{opt}} \quad (5.21)$$

The center of the noise circle (Γ_{noise}) is calculated from [3]:

$$\Gamma_{noise} = \frac{\Gamma_m}{1 + N_{ci}} \quad (5.22)$$

The radius of the noise circle (r_{noise}) is calculated from [3]:

$$r_{noise} = \frac{\sqrt{N^2_{ci} + N_{ci}\left(1 - \left|\Gamma_m\right|^2\right)}}{1 + N_{ci}} \quad (5.23)$$

Noise circles are plotted in Figure 5.7 using the Keysight ADS Smith chart utility for $\Gamma_{opt} = 0.5 < 135°$, $NF_{min} = 0.25$, and $R_e = 0.5$.

5.3.2 LNA Design

LNA requirements vary greatly depending on the type of radar used. Some factors, including size, weight, and cost, are universal to all components. Modern military radars that use AESAs may use hundreds or thousands of LNA elements so a strong emphasis is placed on power consumption. Alternatively, automotive radar places very little emphasis on power consumption since it is small in comparison with other vehicle systems.

Designing an LNA follows the same procedure as designing a power amplifier. The optimal bias is determined by trading off gain and noise (discussed in 5.3.1). Then, the optimal impedance is determined by plotting performance circles on a Smith chart. Figure 5.8 shows a set of source and load curves plotted using ADS.

In Figure 5.8, on the source side (left figure), the region of instability is plotted to ensure that the impedance is stable. Several noise and gain circles are plotted so that a designer can explore the trade space. As shown in Figure 5.8, available gain increases as the impedance moves away from NF_{min}. However,

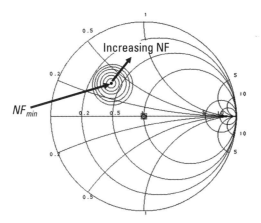

Figure 5.7 Example noise circles with $\Gamma_{opt} = 0.5 < 135°$, $NF_{min} = 0.25$, and $R_e = 0.5$.

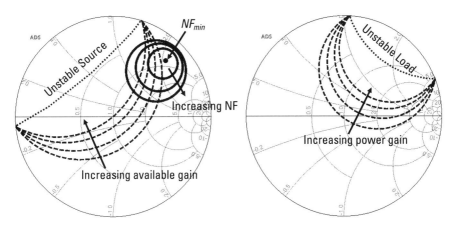

Figure 5.8 Example source (left) and load (curves) used for designing an LNA.

there is a range of impedances within the first noise circle (the one closest to NF_{min}) that touches the second highest gain circle. That is the optimal point for best overall balance.

On the load side (on the right of Figure 5.8), the region of instability is also plotted to ensure the impedance is stable. The load impedance is generally selected for maximum power gain (particularly if this is the first LNA in a chain) or VSWR (not plotted).

5.3.3   Self-Bias Scheme

The biasing strategies discussed have all required separate gate and drain bias voltages to set the desired drain current level. However, LNAs have the option of using a self-bias circuit, which only requires a single positive supply. Figure 5.9 shows a self-bias circuit.

A resistor (R_{bias}) is added between the transistor source and ground. When drain voltage (V_d) is applied, a current (I_{ds}) flows from the drain to the source through the resistor. This causes a voltage drop at the transistor source equal to $I_{ds} \cdot R_{bias}$. If the transistor gate is DC-grounded, then it will be at a lower voltage than the transistor source. Effectively, a negative gate-source voltage (V_{gs}) has been applied. The value of R_{bias} can be set to provide the desired V_{gs}.

In order to work properly, a bypass capacitor must be added to provide the transistor with RF ground at the source. The capacitor value should be chosen to have low impedance at the operating frequency. Additionally, to prevent shorting the RF to ground at the transistor gate, an RF-blocking inductor must be added. The inductor value should be chosen to have high impedance at the operating frequency. Since the RF-blocking inductor and DC-blocking capacitor have a bandwidth-limiting effect, self-biasing is not as broadband as traditional biasing.

Since I_{ds} must flow through R_{bias}, the gain and efficiency of the transistor is reduced. For this reason, self-biasing is better suited for LNAs than power amplifiers.

5.3.4   Gain Equalizers

Gain equalizers or flatteners are used to compensate for an amplifier's natural 6-dB/octave drop in gain over frequency. An equalized amplifier would have the same gain over the operating frequency range. Unequalized gain can lead to signal distortions that can produce suboptimal results and artifacts in radar returns.

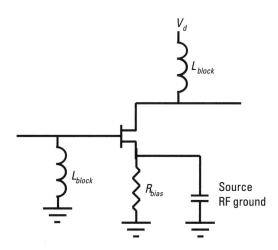

Figure 5.9   Self-biasing circuit.

Equalizers operate by having peak loss at low frequency, minimal loss at high frequency, and a desired loss slope in between. They are created by modifying an attenuator (discussed in Chapter 6) so that series resistors are in parallel with a capacitor and shunt resistors are in series with an inductor. At low frequency, capacitors are high-impedance and inductors are low-impedance so the resistor values are unchanged and attenuation is highest. As the frequency increases, the capacitor impedance drops and the inductor impedance rises to dampen the resistor values. This reduces the level of attenuation. Ideally at high frequency, the capacitors would short circuit and the inductors would open circuit the resistors to eliminate all attenuation. In practice, there will always be some loss. Figure 5.10 presents three examples of equalizer circuits.

Unfortunately, in all cases, resistors are added, which increases the NF of the LNA. The first two circuits are generally more popular for power amplifiers and should always be placed before the amplifier. Otherwise the RLC components must be sized to handle the high output power, and the amplifier efficiency will be reduced by the equalizer losses.

The rightmost circuit in Figure 5.10 can be used to equalize a power amplifier, but it is more popular with LNAs. It is a simple RLC circuit that provides feedback from the output to the input. The capacitor provides a DC block so that gate and drain voltages are isolated. At low frequency, the inductor is low-impedance so the level of feedback is maximized. At high frequency, the inductor is high-impedance so feedback is minimal. Since the magnitude of the output waveform is higher than the input, the resistor dampens the feedback amplitude so the right amount of cancelation is achieved.

To be effective, the RLC circuit must be placed as close to the amplifier transistor as possible. Otherwise, a phase delay will be introduced and feedback cancelation will not happen. This can be challenging, especially for packaged transistors (discussed in Chapter 7).

Regardless of the circuit topology used to equalize performance, it is important to minimize the resistor values by only compensating for the gain slope

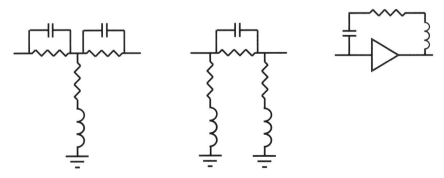

Figure 5.10   Equalizer types are shown: tee (left), pi (center), and feedback (right).

needed. For example, if a specification allows for 2-dB/octave gain slope, less resistance is needed in the equalizer and thermal noise is reduced by nearly 18%.

5.3.5 Resistor Component Selection

Ideally, resistors would be eliminated from the receiver chain (including the LNA) to minimize thermal noise. Unfortunately, that usually isn't possible, but some resistors are better for noise applications than others so proper resistor selection can reduce NF [5].

Carbon composition resistors are made by compressing a mixture of carbon dust or graphite paste with a nonconducting powder. The ratio of conducting to nonconducting material determines the resistor value. Since they have low series inductance and are inexpensive to make, they are very popular for RF applications. They are comprised of compressed particles, so they have flicker noise in addition to thermal noise. At high temperature, their noise performance degrades significantly. In total, they have the highest noise level of any resistor type.

Film resistors also have a layered configuration, but they use well controlled films that are laser-trimmed to determine the resistance value. This allows them to have much finer tolerances than composition resistors. Film resistors can be classified as *thin-film* or *thick-film* depending on the thickness of the conductive material. They are also known as *Cermet resistors* because they integrate CERamic with METal. Since the material is more homogeneous than composition resistors, flicker noise is less. They also degrade less at high temperature.

Wire-wound resistors are made by wrapping a thin metal alloy wire around an insulating ceramic, plastic, or fiberglass rod. Nichrome is typically used as the metal alloy. Compared to the prior resistor types, wire-wound resistors have no flicker noise, so they have the lowest noise. They also handle power and temperature very well. When choosing a wire-wound resistor, it is important to verify that an alternating winding pattern was used. Otherwise, the resistor will have a strong series inductance.

Practical Note

When multiple resistors that seem to meet the requirements are available, choose the one with the lowest power rating. Often, the changes made to increase power handling capability also increase noise. Similarly, choose a thin-film resistor over a thick-film resistor for the lowest noise.

5.4 High Dynamic Range

It would seem that given enough gain stages, a signal—no matter how weak—could be amplified to any desired level. Unfortunately, there is always a limit to the smallest signal that can be amplified, and that limit is determined by noise. Figure 5.11 demonstrates what a signal looks like as the SNR degrades from ∞:1 (noiseless) to 0.25:1 (signal level is four times lower than noise level).

Similarly, there is a maximum power level that the LNA can receive and still operate. The dynamic range of an LNA is defined as the difference in signal amplitude between the maximum linear power level (output is proportional to the input) and the minimum detectible power level (usually determined by noise level, low-level distortion, interference, or resolution level). This is shown in the Figure 5.12 power curve.

In the receiver and signal processor chain of a radar system, the minimum acceptable signal level may be set by the required detection probability (P_d) or tracking accuracy. The minimum acceptable signal level at the processor output is generally 10 or more decibels above the average noise power. However, at the receiver output it may be far below the noise power, as the system relies on subsequent processing gain to achieve the high SNR required for detection or tracking. One must distinguish between LNA dynamic range and the combined receiver and signal processor dynamic range.

Linear power is usually defined as rated output power with an associated IM3 or OIP3 level (discussed in Chapter 4) or at a specific compression level (usually 1-dB compression, or P1dB). Techniques for increasing linear power (and in the process increasing dynamic range) include the following:

- Incorporating a linearizer: Analog and digital linearization techniques are discussed in Section 4.4.3. Since adding resistive elements may be required, it is important to evaluate the impact on NF.

- Biasing to higher drain current: Biasing closer to class A increases the maximum linear power, but it also increases drain current. Figure 5.6 shows that NF_{min} occurs at low drain current. A trade-off can be made between dynamic range and NF.

- Output-matching for high OIP3: Usually the output termination of an LNA is set to achieve maximum gain. In a cascade, this provides the lowest overall NF. However, higher dynamic range can be achieved by terminating at an impedance that trades gain for higher OIP3.

- Using a more linear transistor: Even within a device technology (i.e., GaN transistors), some foundry devices are more linear than others. Some foundries even offer a special linear process that trades off power density, voltage breakdown, or other parameters. A more linear device reduces the difference between P_{sat} and max linear power.

- Using envelope termination [7]: When the signal input is comprised of multiple tones ($f_1, f_2, \dots f_N$) a set of mixing products are generated at large signal strength. If the tone spacing is close, some of these mix-

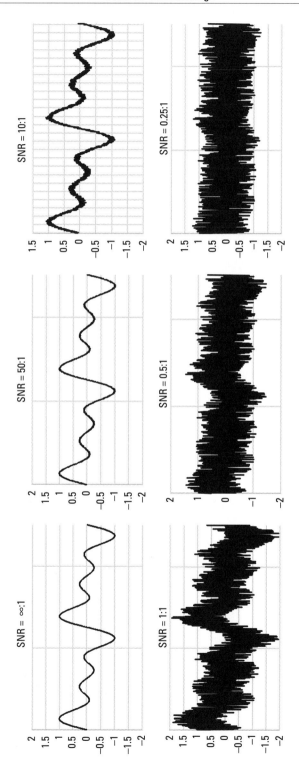

Figure 5.11 Sample waveform as the SNR decreases from ∞:1 to 0.25:1.

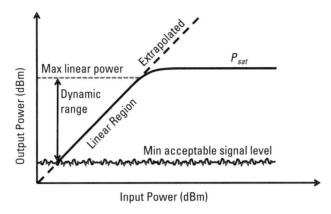

Figure 5.12 Power curve demonstrating definition of dynamic range.

ing products will fall well below band (i.e., $f_2 - f_1$). These products can be eliminated by selecting bypassing capacitor values that filter those frequencies. This technique is not applicable when tone spacing can be wide.

Since the LNA often has a much wider bandwidth than that of subsequent receiver stages, the noise floor at the LNA is likely much higher than that of the receiver. Therefore, the dynamic range can also be extended by limiting the bandwidth of the LNA.

5.5 Cryogenic Operation

There is an increasing interest in operating receivers at or below the cryogenic temperature (defined by the National Institute of Standards and Technology as −180°C, −292°F, or 93.15K). Advancements in microchillers have made cryogenic operation practical, particularly for military applications. Since thermal noise is a significant contributor to overall system noise, reducing operating temperature greatly improves radar receiver sensitivity (especially when external noise sources are low).

Electrical performance generally improves as temperature drops (i.e., gain and efficiency increase and loss is reduced). Designing an LNA for cryogenic operation requires additional considerations, including the following:

- Stability analysis with added gain: A two-stage LNA will have approximately 6.5 dB more gain at −180°C than at 25°C (0.016 dB per stage per °C). Increasing S_{21} by this amount can be sufficient to cause instability. This can be avoided by designing for stability at low temperature.

- Mechanical effects: The mechanical stress added by a change in temperature can be calculated from (4.98). Carrier materials and die attachment methods must be chosen to handle this added stress. Materials must also be chosen for temperature and mechanical stability at low temperature.

- Electrical effects on passive components: Lumped elements and their models have a fixed operating range. Commercial components are usually not tested or rated below −50°C. Component values and parasitic behavior can change at extreme temperatures, which can lead to an unexpected degradation in performance.

5.6   Limiter Elimination

LNAs are designed to be sensitive over a high dynamic range, but the input power level is very low compared to power amplifiers (i.e., microwatt or milliwatt range). Even a small pulse (or burst) of power from a neighboring transmitter can be sufficient to overpower and destroy an LNA. Consequently, a limiter circuit is usually added between the antenna and LNA to protect the LNA from overpower conditions. The limiter is designed to be linear over the dynamic range of the LNA and reach P_{sat} within the power handling range of the LNA. In the event of a burst, the limiter circuit will saturate and prevent excess power from reaching the LNA.

Although effective, limiters add size, weight, cost, loss, and noise to the receiver. The introduction of an LNA that could handle a fair amount of input power level could eliminate altogether. This is not a new concept, but the proliferation of GaN has reenergized this topic. GaN can achieve low noise performance and has four to 20 times higher power handling capability than traditional LNA technologies (i.e., GaAs and InP).

When designing an LNA to handle high input power, the transistor size must be chosen accordingly. For example, an LNA that needs to handle 10-W continuous input power will need 2 mm of gate periphery if the power handling capability is 5 W/mm (typical power density for GaN). In the case of GaN, biasing the drain voltage below the normal operating voltage (i.e., 28V or 50V) usually results in better noise performance. GaN LNAs usually operate in the 5–10V range.

Designing an LNA to handle high power will always result in higher NF. However, if the added noise is equal to or less than the noise contribution from the limiter, it is still a worthwhile venture since size, weight, and cost will be reduced.

Exercises

1. What is the noise temperature of an LNA with 1.5-dB NF?

2. Determine the thermal noise voltage of a 49-Ω, 100-MHz source with 35% bandwidth at 50°C.

3. Convert 3 dB NF to noise factor.

4. The highest temperature ever recorded on Earth is 134°F on July 10, 1913, at Greenland Ranch, Death Valley, California. Likewise, the coldest temperature ever recorded of an inhabited area is –90°F on January 1, 1933, in Oymyakon, Russia. For a given level of noise power, what is the difference in bandwidth for a circuit operating in these two locations?

5. Which amplifier will experience more mechanical stress:
   - An LNA that must operate from cryogenic temperature to 85°C on a brass carrier;
   - A power amplifier that must operate from –20°C to 85°C on an aluminum carrier.

6. A 0.5 W LNA with 60-dB dynamic range operates at 2.5 dB backed off from saturation. What is the minimum acceptable signal level?

7. Use Keysight ADS to plot noise circles for a transistor with default values. Alternatively, perform the same analysis using a transistor model from a commercial foundry.

References

[1] Pozar, D., *Microwave Engineering,* New York, NY: John Wiley & Sons, 1997.

[2] Cohen, L., "The History of Noise," *IEEE Signal Processing Magazine,* Vol. 22, No. 6, Nov. 2005.

[3] Collin, R., *Foundations for Microwave Engineering,* New York, NY: IEEE Press, 2001.

[4] Leach, W., "Fundamentals of Low-Noise Analog Circuit Design," *Proceedings of the IEEE,* Vol. 82, No. 10, Oct. 1994.

[5] Ott, H., *Noise Reduction Techniques in Electronic Systems,* New York, NY: John Wiley & Sons, Inc., 1988.

[6] Senturia, S., *Microsystem Design,* Norwell, MA: Kluwer Academic Publishers, 1940.

[7] Lucek, J., and R. Damen, "Designing an LNA for a CDMA Front End," *RF Design,* Feb. 1999.

Selected Bibiligraphy

Maas, S., *Practical Microwave Circuits,* Norwood, MA: Artech House, 2014.

Bronckers, S., et al., *Substrate Noise Coupling in Analog/RF Circuits,* Norwood, MA: Artech House, 2010.

Maas, S., *Noise in Linear and Nonlinear Circuits,* Norwood, MA: Artech House, 2005.

6

Passive Circuitry

Amplifiers, like rock stars, seem to get all the attention, but neither could achieve greatness without support. Rock stars could not take center stage without an army of support staff. Similarly, amplifiers rely on surrounding passive circuit elements to make them better. Passive circuit components can reduce or eliminate unwanted frequencies generated from amplifier nonlinearity. Moreover, they can compensate for impedance mismatch between components, and they are responsible for sending and receiving energy. These are only a few simple examples of what passive elements bring to the radar system.

Designed properly, passive circuits can take a radar front end to a new level. Designed poorly, passive circuits can really hurt performance. They can limit bandwidth and power handling. They always add loss and size, but if those are excessive, they can place an unnecessary burden on the active circuits. The parasitics they introduce can reduce radar performance.

This chapter discusses design strategies for the most common passive circuit components used in radars, reviewing methods to minimize layout discontinuities and analyze signal flow through a circuit and presenting inherent limitations to bandwidth and performance.

6.1 Limiting Factors and Ways to Mitigate

Before reviewing the design procedure for individual components, we present general issues that affect all passive circuits.

6.1.1   Lumped Elements

Misconceptions about lumped elements (i.e., surface-mount components) are described as follows.

- Self-containment (no radiation): Components are rarely provided with shielding, so they will radiate (discussed in Chapter 8). Poorly placed components will couple and affect isolation and linearity.

- Parasitics: At RF frequencies, all components have parasitics (unwanted resistance, capacitance, and inductance that arise due to the nature of the materials used to implement them) that can result in loss and changes in the circuit frequency response. Chapter 3 presents equivalent circuit models. Even the highest-quality lumped elements available have nonideal behavior at some frequency.

- Equal performance over rated band: Particularly in wideband applications, the performance of a component will change over frequency. Data sheets generally choose the center frequency performance to be the nominal condition. So, part of the rated band will be above that number and part will be below. For example, a nominal 5-nH inductor may be 4.5 nH at the low end of the frequency band and 5.5 nH at the high end.

- Loss: All of the equivalent circuit models presented in Chapter 3 include a resistive component, so they all have loss. This is often forgotten, especially for purely reactive elements such as capacitors and inductors. A reactive-element matching circuit containing multiple sections will inherently have loss from parasitic resistance and from the transmission lines that connect these sections.

- Smaller component size supports operation at higher frequency: It is important that the physical size be less than (and ideally much less than) the wavelength of the highest operating frequency. In some cases, however, the small size of a lumped component results from the need to reduce cost, rather than design for high-frequency performance. So it is important to verify the performance as well as the size before the component is used in a high-frequency circuit.

6.1.2   Bode-Fano Limit

As discussed in Chapter 4, there are trade-offs between impedance match and other performance metrics (i.e., output power and efficiency). Sometimes impedance match must be sacrificed to meet other objectives. When it comes to achieving a good impedance match over a broad bandwidth, there is a fundamental limitation that was determined in the 1940s and 1950s by Bode and Fano [1]. The Bode-Fano, which limit relates the reflection coefficient to a load impedance, can be used to determine the maximum bandwidth obtainable for a given quality match. Every impedance $(R+jX)$ has an equivalent circuit comprised of a series or shunt RC or RL circuit. The Bode-Fano bandwidth-limiting equations are listed as follows, with $\Gamma(\omega)$ as the reflection coefficient of the load:

• Shunt RC:

$$\int_0^\infty \ln\frac{1}{|\Gamma(\omega)|}d\omega < \frac{\pi}{RC} \tag{6.1}$$

• Series RC:

$$\int_0^\infty \ln\frac{1}{|\Gamma(\omega)|}d\omega < \pi\omega_0^2 RC \tag{6.2}$$

• Shunt RL:

$$\int_0^\infty \ln\frac{1}{|\Gamma(\omega)|}d\omega < \frac{\pi\omega_0^2 L}{R} \tag{6.3}$$

• Series RL:

$$\int_0^\infty \ln\frac{1}{|\Gamma(\omega)|}d\omega < \frac{\pi R}{L} \tag{6.4}$$

In order to assess the meaning of (6.1)–(6.4), we can define the load as Z_m and $|\Gamma|$ as shown in Figure 6.1.

In this case, we can simplify the equations to:

$$\int_0^\infty \ln\frac{1}{|\Gamma(\omega)|}d\omega = \int_{\Delta\omega} \ln\frac{1}{\Gamma_{min}}d\omega = \Delta\omega\ln\frac{1}{\Gamma_{min}} < Z_m \tag{6.5}$$

A plot of $\Delta\omega$ versus Γ_{min} for several values of Z_m is shown in Figure 6.2. This plot allows us to draw some important conclusions:

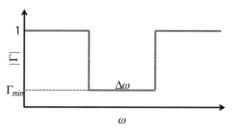

Figure 6.1 Definition of |Γ| for simplifying (6.1)–(6.4).

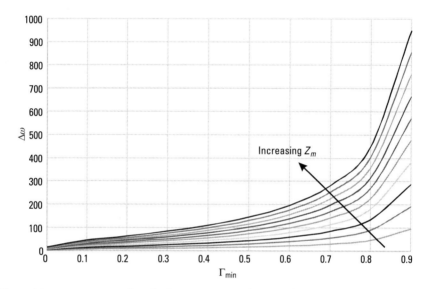

Figure 6.2 Δω versus Γ_{min} from (6.5) with varying Z_m (10 to 100).

- Regardless of the load (Z_m), as the bandwidth (Δω) increases, the reflection coefficient (Γ_{min}) must also increase.

- A perfect match ($\Gamma_{min} = 0$) can only be achieved at one frequency (Δω = 0).

- A near perfect mismatch ($\Gamma_{min} > 0.9$) has an indeterminate bandwidth (the limit approaches infinity as Γ_{min} approaches 1).

- At a fixed bandwidth, Γ_{min} will decrease as Z_m increases (in the case of a parallel RC load, that would require a decrease in the RC value).

Practical Note

The Bode-Fano equations proved without exception that very broadband applications will not achieve a good instantaneous impedance match. In practice, the theoretical bandwidth limit set by the Bode-Fano equations cannot be achieved. However, understanding this limit is important to avoid chasing an impossible specification.

6.1.3 Discontinuities

A discontinuity occurs whenever a transmission line ceases to be straight and perfectly continuous. This includes bends, gaps, open circuits, short circuits, width changes, material changes, external loadings (i.e., shunt stubs), and poor connections. These changes can be modeled using lumped elements and can be beneficial or detrimental to performance. For example, discontinuities can be used for filtering or impedance-matching. They can also contribute unwanted parasitics.

A common discontinuity is the right-angle bend. To keep layouts small, transmission lines are meandered like an accordion to increase electrical length while minimizing physical size. Figure 6.3 shows three examples of a right-angle bend: The leftmost image shows a cornered bend with the most parasitic inductance; the center image is mitered and has reduced parasitic inductance; and the rightmost image shows a rounded bend with the lowest parasitic inductance.

Although rounded bends have the best performance, they are also the most difficult (and therefore expensive) to produce. For this reason, mitered bends are the most widely used. They are implemented by removing the corner, as shown in Figure 6.4.

A design equation for a mitered right-angle bend is available for $0.25 \leq w/h \leq 2.75$ and $2.5 \leq \varepsilon_r \leq 25$ [2]:

$$a = w\left(1.04 + 1.30e^{-135\frac{w}{h}}\right) \tag{6.6}$$

where a is the dimension shown in Figure 6.4, w is the signal line width (meters), and h is the substrate height (meters). The calculated value for a is not always the optimal value, but it is usually close.

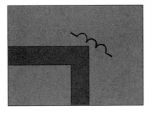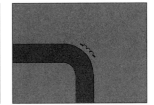

Figure 6.3 Three types of right-angle bends with their magnitudes of parasitic inductance shown graphically.

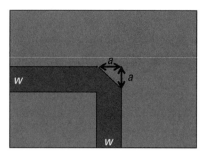

Figure 6.4 Dimensioned layout for optimal miter in right-angle bend.

The most accurate way to determine the optimal miter is through full-wave simulation (discussed in Chapter 8). Table 6.1 shows full-wave simulated results of return loss, insertion loss, and change in insertion phase for a 50-Ω (25 mil) mitered bend on 25-mil-thick alumina ($\varepsilon_r = 9.6$). The optimal miter is 35 mil. For comparison, the last row is for a rounded bend. The rounded bend loss is lower than the optimal miter, but the additional electrical length increases the phase response.

A similar perturbation can be made wherever there is a tee (or T) junction to reduce parasitic effects, as shown in Figure 6.5.

The parasitic inductance can be reduced by removing part of the transmission line as shown in Figure 6.5 (right image). As the signal propagates down the line, the improved tee-junction makes the split much less abrupt, so current *hot spots* are mitigated. This reduces the chance of undesired radiation.

Table 6.1

S_{11}, S_{21}, and Change in Insertion Phase for Varying Miter Values of a Right-Angle Bend and Rounded Bend (*a* Defined in Figure 6.4)

a	S_{11} (dB)	S_{21} (dB)	$\Delta\angle S_{21}$
0 mil	−15.6	−0.159	0
5 mil	−15.9	−0.151	0.295
10 mil	−16.5	−0.135	0.898
15 mil	−17.5	−0.114	1.706
20 mil	−19.1	−0.089	2.661
25 mil	−21.7	−0.065	3.712
30 mil	−26.8	−0.043	−3.169
35 mil	−49.9	−0.033	−10.1
40 mil	−25.3	−0.045	−17.2
45 mil	−18.2	−0.098	−24.617
47.5 mil	−15.6	−0.154	−28.6
Rounded	−38.6	−0.025	−109

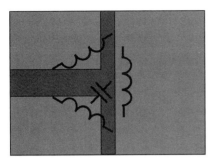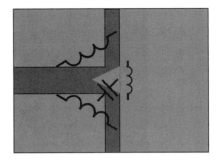

Figure 6.5 Original (left) and improved (right) tee-junction design to reduce parasitic inductance.

The optimal shape of the perturbation is best determined by full-wave simulation, but it is common for the inset to be 0.5 to three times the higher-impedance line width.

Often, metal traces are added to serve as landing pads for optional tuning elements. Figure 6.6 shows an example of such a trace.

These traces are effectively parallel-plate capacitors in disguise, and their value should be calculated to ensure they are sufficiently low (or compensated for). The design equation is [2]:

$$C(\text{pF}) = \frac{8.8542 \cdot \varepsilon_r}{1000} \frac{(w-h)(l-h)}{h} + \frac{26.40(\varepsilon_r + 1.41)}{1000} \frac{w+l}{\ln\left(\frac{5.98h}{0.8h+t}\right)} \quad (6.7)$$

where all dimensions were shown in Figure 6.6, all lengths are in millimeters, $w > h$, and $l > h$.

Figure 6.7 shows the equivalent capacitance as the trace area $(w \cdot l)$ and substrate height (h) increase. The figure includes plots with permittivity 2 (left

Figure 6.6 Dimensioned layout of metal trace on substrate.

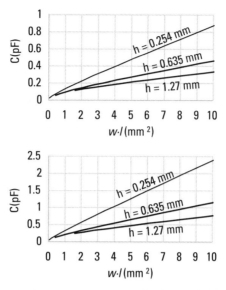

Figure 6.7 Equivalent capacitance as trace area and substrate height increase on materials of permittivity 2 (left) and 6 (right).

image) and 6 (right image). For thin substrates (10–15 mil) and moderately high permittivity (6+), the effective capacitance is nonnegligible.

6.2 Couplers

Directional couplers are four-port components that can split or combine a signal. The four ports are labeled on two different (but equivalent) component symbols shown in Figure 6.8.

The through (P_2) and coupled (P_3) ports are the intended outputs. If the outputs are meant to be equal amplitude (3-dB coupler), then $S_{21} = S_{31} = -3$ dB (plus any losses in the component). If the outputs are unequal, then the coupling factor (C) is defined by:

$$C = 10\log\frac{P_1}{P_3} = 20\log\frac{1}{|S_{31}|} \tag{6.8}$$

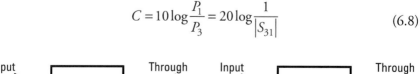

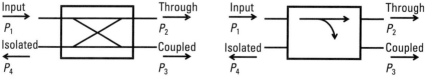

Figure 6.8 Equivalent symbols for directional couplers with labeled ports.

where P_1 is the power at port 1 (W) and P_3 is the power at port 3 (W).

In a perfect coupler, the Isolated (P_4) port would have no output signal. Since no real component is perfect, there are two metrics for assessing coupler quality. Directivity (D) relates the output power to the coupled port and the isolated port and can be calculated by:

$$D = 10\log\frac{P_3}{P_4} = 20\log\left|\frac{S_{31}}{S_{41}}\right| \tag{6.9}$$

Isolation (I) relates the *leakage* from the input port to the isolated port and can be calculated by:

$$I = 10\log\frac{P_1}{P_4} = 20\log\frac{1}{|S_{41}|} \tag{6.10}$$

In a perfect coupler, both the directivity and isolation would be infinite. The above quantities are related by:

$$C = I - D\,(\text{dB}) \tag{6.11}$$

In a quadrature or 90-degree hybrid design, the outputs (ports 2 and 3) have a 90-degree phase difference between them. The most common quadrature design uses a branch-line strategy as shown in Figure 6.9. The same design can be implemented in a circular form where each side of the square is 90 degrees.

An equivalent circuit model is also shown in Figure 6.9 (right image). The component values can be determined by [3]:

$$L_1 = \frac{Z_o}{2\sqrt{2}\pi f} \tag{6.12}$$

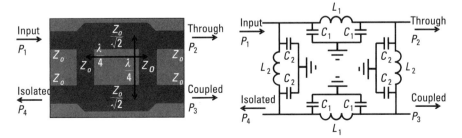

Figure 6.9 Microstrip layout and equivalent circuit model for quadrature coupler.

$$C_1 = \frac{\sqrt{2}}{2\pi f Z_o} \tag{6.13}$$

$$L_2 = \frac{Z_o}{2\pi f} \tag{6.14}$$

$$C_2 = \frac{1}{2\pi f Z_o} \tag{6.15}$$

To achieve broadband coupling, multiple hybrid components can be cascaded. Alternatively, a 90-degree Lange coupler design can be used. It is a type of quadrature coupler with many closely spaced parallel lines. This architecture is also advantageous when high coupling factors are needed. Figure 6.10 shows an example topology. Unfortunately, these designs can be difficult to fabricate because of the tight tolerances required and the need for tightly controlled wire bonds or air bridges. However, they can be made very compact and broadband. Section 7.5.1 presents an example of this architecture.

In a 180-degree hybrid, the outputs have 180 degrees phase difference between them. Figure 6.11 shows an example ring hybrid or rat race layout and the equivalent circuit model. There are a number of ways it can be operated, described as follows.

- When P_1 is the input, P_2 and P_3 are in-phase outputs (90 degrees shifted from the input) and P_4 is isolated;

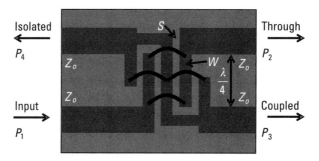

Figure 6.10 Example layout of Lange coupler.

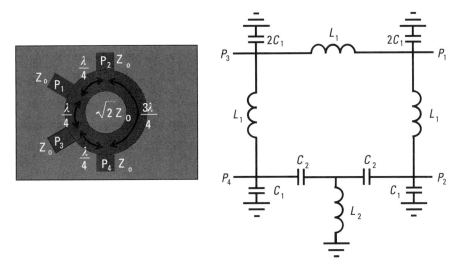

Figure 6.11 A 180-degree hybrid layout and the equivalent circuit model.

- When P_4 is the input, P_2 and P_3 are 180-degree out-of-phase outputs and P_1 is isolated;
- When P_2 and P_3 are the inputs (as a combiner), P_1 is the summing port and P_4 is the difference port.

The component values can be determined by [3]:

$$L_1 = \frac{\sqrt{2}Z_o}{2\pi f} \tag{6.16}$$

$$C_1 = \frac{1}{2\sqrt{2}\pi f Z_o} \tag{6.17}$$

$$L_2 = \frac{\sqrt{2}Z_o}{2\pi f} \tag{6.18}$$

$$C_2 = \frac{1}{2\sqrt{2}\pi f Z_o} \tag{6.19}$$

Practical Note

Directional couplers are commonly used in measurement benches to determine incident and reflected power; directional couplers are also known as "reflectometers." When connected in line, the input signal can be measured at the coupled port (P_3) and the reflected power can be measured at the Isolated port (P4).

6.3   Isolators and Circulators

Isolators and circulators are two-port and three-port (respectively) components that force electromagnetic energy to propagate in one direction. This is achieved by the use of ferrites. Whenever possible, these components are generally avoided for several reasons:

- Ferrites are made from iron cores, which are heavy for their size (an issue for airborne platforms).

- Little can be done to significantly shrink the size of ferrites, so components tend to be large (an issue meeting new size constraints).

- Manufacturing ferrites is a time-consuming and expensive process, which drives the high cost of isolators and circulators (not a low-cost solution).

- Ferrites are inherently narrowband. Generally wideband ferrites are an octave in bandwidth (suitable for half-octave radars, but not ultra-wideband radars).

- Ferrites are nonlinear components, so if overdriven, they will degrade in linearity and can even fail.

Practical Note

Isolators are just circulators with one port terminated. Often, circulators are more readily available off the shelf. If all you need is an isolator, consider purchasing a circulator and a 50-Ω load to hasten the procurement cycle and lower the cost.

When isolators and circulators are not practical, they can be replaced with couplers (discussed in Section 6.2) by taking advantage of the isolation port or with switches to toggle between transmit and receive modes.

6.4 Switches

Switches are used in nearly all electronics to control the flow of a signal. In a radar, they can be used to toggle between transmit and receive circuitry. In a phased array, they are used to shift phase (Section 6.5) and attenuation (Section 6.6).

Switches have four main parameters:

- Number of inputs (poles) and number of outputs (throws);
- Insertion loss or ON resistance (ideally zero);
- Isolation or OFF capacitance between input and output ports (ideally infinite);
- Reflective type (the OFF port is open or short circuit) or nonreflective/ absorptive/ terminated/ resistive type (the OFF port is set to the system impedance, usually 50Ω);
- An equivalent circuit model for an ON and OFF switch is shown in Figure 6.12. C_p and R_p are generic parasitic elements [3].

Unfortunately, due to the wide variety of switching technologies available, universal design equations for the circuit elements are not possible. These values are best determined empirically by fitting circuit performance to measured data.

For RF applications, solid-state switches are principally used instead of mechanical switches (i.e., relays) due to their faster turn-on/off speed, smaller size, and lower loss. In recent years, microelectrical-mechanical switches (MEMS) have gained in popularity in radar applications due to their broad bandwidth, excellent linearity, good isolation, and increasing power handling capability. What limits their widespread usage is increased cost and size, difficulty in integration, lower reliability, shorter lifetime, slower switching speed (microseconds versus nanoseconds), lower temperature handling (less than 100°C versus more than 200°C), and high switching voltage (up to 100V instead of less than 5V)

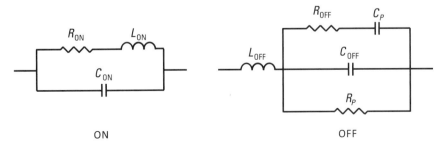

ON   OFF

Figure 6.12 Equivalent circuit model for an ON and OFF switch.

compared to solid-state switches. However, progress is being made in all of these areas.

In 2006, Radant MEMS, Inc., (Stow, MA) demonstrated an electronically scanning antenna for air and surface target detection using 25,000 MEMS switches with a scanning capability of ±60° and operated over a 1-GHz bandwidth at X-band. This is believed to be the world's first demonstration of a MEMS-based radar system [4]. Unfortunately due to the inherent "mechanical" nature of MEMS, reliability and lifespan continue to be issues impeding their universal adoption.

Solid-state switches can be made from diodes (i.e., PIN switches) or transistors (i.e., FET switches). PIN switches are popular due to their high-power handling and low cost. However, they are more cumbersome to use because they require a current flow to turn ON. An example bias network is shown in Figure 6.13. The PIN switch is represented symbolically as a diode.

DC-blocking capacitors are used at the input and output of the PIN switch to isolate current flow to that component. RF-choking inductors are added to prevent the signal from propagating into the voltage source or the ground return. Sections 4.1, 4.5, and 4.6 discuss design rules for DC blocks and RF chokes using (4.5) and (4.6).

FET switches are easier to use because they have dedicated low-current control (or enable) ports. They are also significantly faster than PIN switches. Figure 6.14 shows example reflective and nonreflective single-pole double-throw (SPDT) FET switches [3].

In Figure 6.14, in the reflective (top) circuit, there is one series and two shunt FETs per throw. The FETs are turned ON (short circuit) and OFF (open circuit) through control voltages (V_{c1} and V_{c2}). To enable output 1, V_{c1} is set to turn ON FETs C, E, and F and V_{c2} is set to turn OFF FETs A, B, and D. The RF signal at the input propagates across C and through output 1. FET D is OFF so that it blocks the signal from propagating through that path. Since no FET can serve as a perfect block, FETs E and F are shorted to ground so any

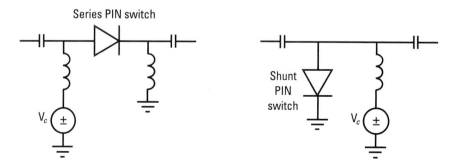

Figure 6.13 Example bias network for a series and shunt PIN switch.

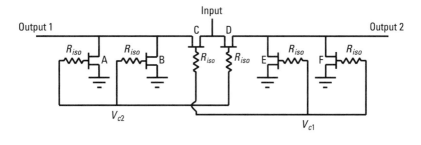

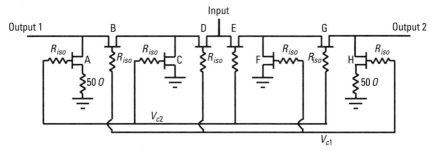

Figure 6.14 Reflective (top) and nonreflective (bottom) SPDT FET switch circuits.

leakage signal is shorted to ground instead of propagating through output 2. Output 2 isolation is improved as additional shunt elements are added. Adding shunt elements improves isolation when that path is OFF, but they add loss when that path is ON. This trade-off drives the compromise between low insertion loss and high isolation. To enable output 2, V_{c1} is set to turn ON FETs A, B, and D and V_{c2} is set to turn OFF FETs C, E, and F. The resistance of the OFF port in this circuit is ideally 0Ω.

In the nonreflective (bottom) circuit in Figure 6.14, there are two series and two shunt FETs per throw. The operation is the same as the reflected case, except the OFF port is connected through an additional FET to 50Ω.

In practice, switches cannot achieve perfect short- and open circuit performance. The ON and OFF resistance of a FET determines the insertion loss and isolation of a switch. R_{on} can range from <10 mΩ to a few ohms, depending on transistor size and technology. R_{off} can range from 1 kΩ to 1 MΩ. To prevent RF from leaking into the bias (gate) port, resistors are added to each FET (R_{iso}). The value of this resistance must be large enough to block RF, but not so large as to slow down the switching speed. Typically values are 1–5 kΩ per millimeter of gate periphery. The control voltage to turn ON a FET (V_{ON}) is usually 0V or slightly higher. A FET is turned OFF (V_{OFF}) by biasing well below pinch-off (< −5V).

High-power radars often use stacked-FETs (or *N-FETs*) to increase power handling. Figure 6.15 shows a stacked-FET circuit for a single shunt configuration.

The RF signal is distributed across the FETs so that the total power handling is increased. Increasing the number of FETs increases the power handling, but it also adds size. The maximum output power can be calculated from [3]:

$$P_{max} = \frac{\left[N\left(V_{BR} - V_P\right)\right]^2}{2Z_o} \qquad (6.20)$$

where N is the number of transistors, V_{BR} is the gate-drain or gate-source breakdown voltage (volts), and V_P is the pinch-off voltage (volts). Since there is a squared relationship, doubling the number of transistors quadruples the power handling.

The operating frequency range of a switch can be improved by adding components to compensate for the parasitic elements shown in Figure 6.12. For example, adding a series inductance between shunt FETs will minimize the parasitic capacitance and extend the frequency range. Using devices with low C_{OFF} (such as small gate length devices) will also improve operating frequency.

If a single negative voltage is not available, bias can be implemented using two positive voltages as shown in Figure 6.16 [3].

A fixed voltage (V_{fixed}) is applied to the source terminal of the FET. The voltage should be set to $|V_{OFF}|$. If the control voltage (V_c) is set to 0V, then $V_{GS} = -V_{OFF}$ and the FET is OFF. If V_c is set to $|V_{OFF}|$, then $V_{GS} = 0$V (the gate and source are at the same voltage potential) and the FET is ON.

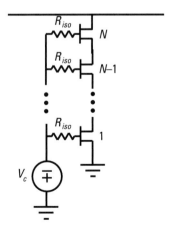

Figure 6.15 Stacked-FET circuit.

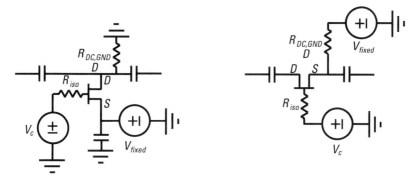

Figure 6.16 Bias network of shunt and series FET using two positive voltages.

6.5 Phase Shifters

All components have electrical length so they exhibit a phase response. If multiple signals are being combined, combining efficiency is maximized when the signals are in-phase. Phase shifters can be added to compensate for phase misalignment by delaying one of the signals until the other is in alignment. In an AESA, changing the relative phase within the antenna array enables electronic beam steering. An example switched-line phase shifter is shown in Figure 6.17. The gaps indicate the location of a switch that enable signal propagation through that path. The l_1 path is enabled by turning ON the top two switches and turning OFF the bottom two. The l_2 path is enabled by turning ON the bottom two switches and turning OFF the top two.

The phases of the l_1 and l_2 paths are:

$$\phi_{l_1} = \frac{360 l_1}{\lambda}(\text{deg}) = \frac{2\pi l_1}{\lambda}(\text{rad}) \tag{6.21}$$

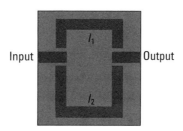

Figure 6.17 Switched-line phase shifter layout.

$$\phi_{l_2} = \frac{360 l_2}{\lambda}(\text{deg}) = \frac{2\pi l_2}{\lambda}(\text{rad}) \qquad (6.22)$$

where l_1 and l_2 are the path lengths (m) and λ is the wavelength (m).

The relative phase shift is:

$$\Delta\varphi = \phi_{l_2} - \phi_{l_1} = \frac{360}{\lambda}(l_2 - l_1)\ (\text{deg}) = \frac{2\pi}{\lambda}(l_2 - l_1)\ (\text{rad}) \qquad (6.23)$$

The time delay of using this approach can be calculated from:

$$\Delta\tau = \frac{1}{f\lambda}(l_2 - l_1) \qquad (6.24)$$

where τ is the time delay (s), f is the frequency (Hz), and λ is the wavelength (m).

The circuit shown in Figure 6.17 is a one-bit phase shifter with two possible phase shifts. More than one of these components can be cascaded to create a multibit phase shifter. The four-bit phase shifter shown in Figure 6.18 is capable of 16 relative phase shifts (0° to 337.5° in 22.5° increments). Compact designs are also possible [5].

6.6   Attenuators

Attenuators are two-port components that add loss (or remove gain) wherever they are placed. When so much effort is spent to reduce loss, it may seem strange that anyone would want to add loss to a system. However, in applications where a broadband match is needed and excess gain is available, adding an

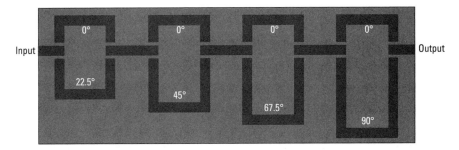

Figure 6.18   Four-bit switched-line phase shifter layout.

attenuator will correct a poor VSWR. A small attenuator at the input or output of an amplifier can maintain stability even when presented with an open or short circuit. Attenuators also serve as broadband loads for terminating isolation ports of couplers or isolators.

The three most popular attenuator circuits are the tee, bridged tee, and pi networks. Figure 6.19 shows all three of these circuits [6].

The design equations are presented as follows, with A as the desired attenuation (dB).

- Tee network:

$$R_1 = Z_{out} \frac{10^{A/20} - 1}{10^{A/20} + 1} \tag{6.25}$$

$$R_2 = 2 Z_{out} \frac{10^{A/20}}{10^{A/10} - 1} \tag{6.26}$$

- Bridged-tee network:

$$R_1 = Z_{out} \left(10^{A/20} - 1\right) \tag{6.27}$$

$$R_2 = \frac{Z_{out}}{10^{A/20} - 1} \tag{6.28}$$

- Pi network:

$$R_1 = Z_{out} \frac{10^{A/20} + 1}{10^{A/20} - 1} \tag{6.29}$$

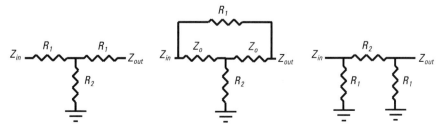

Figure 6.19 Tee, bridged tee, and pi attenuator circuits.

$$R_2 = \frac{Z_{out}}{2} \frac{10^{A/10} - 1}{10^{A/20}} \tag{6.30}$$

Generally, the most appropriate circuit is the one with the most convenient resistor values for the attenuation needed. Figure 6.20 shows values for R_1 and R_2 as the attenuation is swept from 0 to 40 dB. Z_{out} is set to 50Ω. Regardless of the circuit approach, very low attenuation levels are difficult to realize due to the very small or very large resistor values needed.

6.7 Filters/Diplexers

Filters are components that attenuate unwanted frequencies (the stopband) and transmit desired frequencies (the passband) with as little loss as possible. There are five types of filters:

- Low-pass: Only low frequencies propagate;

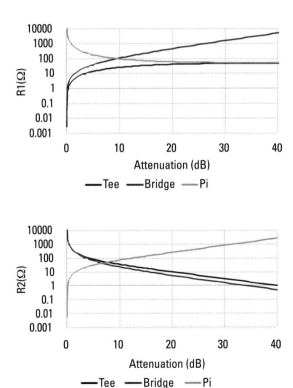

Figure 6.20 R_1 and R_2 as attenuation is swept from 0 to 40 dB.

- High-pass: Only high frequencies propagate;
- Bandpass: Only a frequency range of interest propagates;
- Bandstop (also known as band reject or notch): All frequencies except a range of interest propagate;
- Multipass (or multistop): Multiple regions of frequencies pass (or are rejected).

Unfortunately, no filter is precise enough to pass one frequency and reject the frequency infinitesimally lower or higher. Instead, filters have *skirts* that can be designed to roll off quickly or slowly. Higher-order filters roll off more quickly than lower-order filters, but they require more elements and size (which adds loss). Additionally, the faster the roll-off, the more ripple that occurs in the passband. A response that has a moderate skirt slope and flat passband is called a *maximally flat* or *Butterworth* response. A response that has a steep skirt slope and an equal-amplitude ripple passband is called a *Chebyshev* (or *Tchebychev*) response. Figure 6.21 illustrates these terms.

Group delay is the time required to propagate through the filter. Having constant group delay allows signals to remain undistorted due to frequency dispersion. Pulses, for example, will maintain their rise and fall times if group delay is small.

Filters are one of the most commonly published RF components, and design strategies are as varied as the authors themselves. Using a simulator

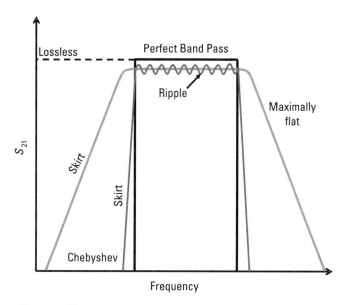

Figure 6.21   Bandpass filter response.

with built-in design guides (discussed in Chapter 8) is really advantageous for determining the best design strategy. Figure 6.22 shows a distributed and lumped-element filter designed using Keysight ADS' Filter and Passive Circuit DesignGuides. The filters provide maximally flat performance from 8 to 12 GHz (all of X-band) with 30-dB rejection 1 GHz out of band (meaning S_{21} at frequencies below 7 GHz and above 13 GHz must be 30 dB below the S_{21} level between 8 and 12 GHz). With this information, the designer can determine whether a distributed or lumped-element approach works best.

6.8   Splitters/Combiners

Power splitters (or dividers) enable a signal to be split into two or more separate paths. Both amplitude and phase can be designed to be equal or unequal at the output. Common applications include feeding networks for an antenna array or a bank of parallel-combined power amplifiers.

Power combiners receive two or more signals. As with splitters, they can be designed to receive equal or unequal amplitude and phase. The most common application is to combine the output power from multiple sources.

The Wilkinson power divider offers isolation at the output ports using a resistor. In the three-port, equal-split configuration (shown in Figure 6.23), all ports are matched to Z_o. This makes it a very versatile design that has led to its widespread use.

Practical Note

Isolation between output ports is important in real applications. If a splitter circuit does not provide isolation (as is the case in a simple tee junction), a change in impedance at any port will affect the impedance at all other ports. This can have a significant impact on circuits that are sensitive to input impedance, such as receivers. At least 10-dB isolation should always be implemented.

The Wilkinson typology supports unequal power ratios as well, as shown in Figure 6.24 [7].

The power ratio between port 2 and port 3 (P_3/P_2) can be calculated from:

$$R_P = \sqrt{\frac{P_3}{P_2}}$$

(6.31)

where P_3 and P_2 are both in linear units (W). The impedance for the two signal paths can be calculated from:

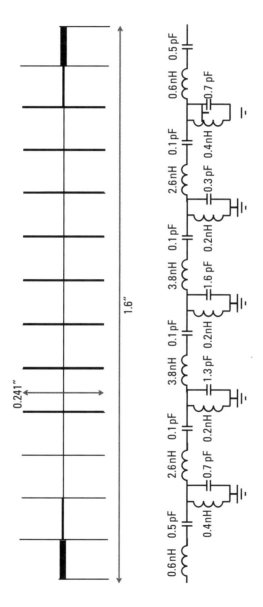

Figure 6.22 Distributed (top) and lumped-element (bottom) bandpass filter.

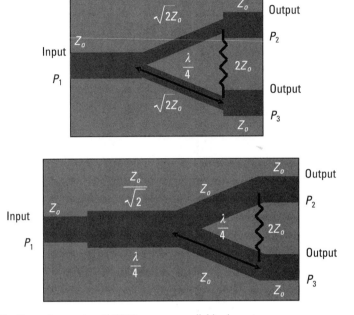

Figure 6.23 Example equal-split Wilkinson power divider layouts.

$$Z_{o,2} = Z_o \sqrt{R_P \left(1 + R_P^2\right)} \tag{6.32}$$

$$Z_{o,3} = Z_o \sqrt{\frac{1 + R_P^2}{R_P^3}} \tag{6.33}$$

The isolation resistor value can be calculated from:

$$R_i = Z_o \left(R_P + \frac{1}{R_P} \right) \tag{6.34}$$

Unlike with the equal-split configuration, the output port impedances are not equal to Z_o. Instead, they can be calculated from:

$$Z_2 = Z_o R_P \tag{6.35}$$

$$Z_3 = \frac{Z_o}{R_P} \tag{6.36}$$

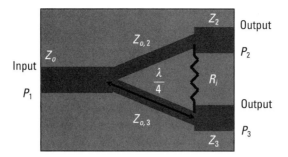

Figure 6.24 Example unequal-split Wilkinson power divider layout.

Notice that the output ports are both multiplied and divided by the square root of the power ratio (R_P). For large power ratios, this can be problematic. For example, if R_P is 10 and the system impedance is 50Ω, the output port impedances will be multiplied and divided by 3.16 (158.11Ω and 15.81Ω respectively). Matching networks will need to be added at port 2 and port 3 to match to 50Ω.

Figure 6.25 shows a circuit that is matched to 50Ω at all ports and incorporates a matching network at the input (Z_1) [6]. The design equations for Figure 6.25, (6.37)–(6.40), are presented as follows.

$$Z_1 = Z_o \left(\frac{R_p}{1+R_p^2} \right)^{0.25} \tag{6.37}$$

$$Z_2 = Z_o R_p^{0.75} \left(1 + R_p^2 \right)^{0.25} \tag{6.38}$$

$$Z_3 = Z_o \frac{\left(1 + R_p^2 \right)^{0.25}}{R_p^{1.25}} \tag{6.39}$$

$$R_i = Z_o \frac{1 + R_p^2}{R_p} \tag{6.40}$$

When the power split ratio is large, the use of high and low impedance signal traces is unavoidable. Some substrate configurations are better suited to realize high and low impedance lines concurrently. Figure 6.26 shows the range

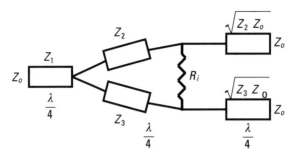

Figure 6.25   Example unequal-split Wilkinson power divider circuit with matched ports.

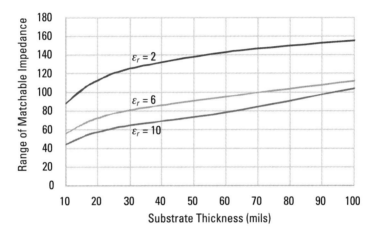

Figure 6.26   Range of matchable impedance values versus substrate thickness for permittivity values 2, 6, and 10.

of matchable impedances versus substrate thickness for three permittivity values. The range of impedance values is defined as the impedance range achievable if the signal width is varied from 10 mils to 250 mils. Figure 6.26 shows that low permittivity and thick substrates are better for matching low and high impedance values.

The Wilkinson topology also supports configurations with more than two outputs. Such components are called *N-way* configurations. Figure 6.27 shows four-way layout. All paths have an impedance of the square root of N multiplied by the system impedance (in the case of Figure 6.27, $\sqrt{4}\,Z_o = 2Z_o$) and are $\lambda/4$ in length. The isolation resistors are equivalent to the system impedance. Notice that all resistors share a common terminal.

The difficulty designing an N-way Wilkinson is with the layout. All paths must have the same physical length and that is challenging to accomplish while maintaining enough distance to prevent coupling. Furthermore, the isolation

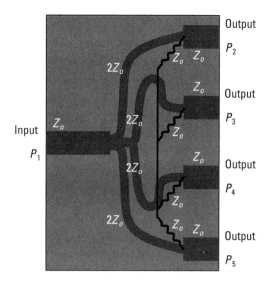

Figure 6.27   A four-way equal-split Wilkinson power divider layout.

resistors must jump over the RF signal traces, and for best results, symmetry should be maintained in all paths. Figure 6.28 shows a layout of a 10-GHz four-way splitter on 25-mil-thick substrate ($\varepsilon_r = 6$). Two wire bonds are needed on the output for isolation.

6.9   Baluns

For DC current to flow in a circuit, there must be a closed loop. Breaking the loop (say, by toggling a switch or removing a series component) will cause DC current to stop flowing. Similarly, when an RF signal propagates down a transmission line, there is a current return. In a single-ended component, such as a microstrip circuit, the current return is through a dedicated ground plane. Since there are two conductors and one is at a lower potential than the other, this is known as an *unbalanced configuration.*

Alternatively, a single-ended component can be implemented in a *differential* configuration. Figures 2.17 and 2.22 present examples of differential transmission lines. In this case, a pair of conductors transmits the signal voltage so neither is dedicated as ground. The two conductors have equal but opposite amplitude, so this is known as a *balanced configuration.* The phase between conductors is 180 degrees.

A balun is a component that converts between an unbalanced and balanced configuration (or vice versa). The name *balun* comes from the words *balanced-unbalanced.* They are commonly used to implement push-pull amplifiers

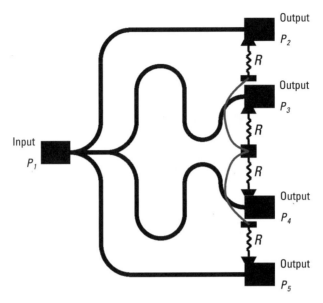

Figure 6.28 Actual layout of 10-GHz four-way splitter on 25-mil-thick substrate with permittivity of 6.

(discussed in Chapter 4.3.3), to feed dipole antennas, and as a power splitter/combiner wherever 180-degree phase shifts between signals can be used.

There are three popular methods for designing baluns for radar. The first utilizes a quarter-wave length of coax line as shown in Figure 6.29. The outer conductor is connected to common ground on the unbalanced end. The center and outer conductors create the balanced end on the other side of the coax. Unfortunately, due to mechanical limitations, coax lines are only practical at lower frequencies.

Planar baluns are popular because they are practical at higher frequencies. Although there are many planar architectures, two common ones are shown in Figure 6.30. The structure on the left of Figure 6.30 looks similar to a CPW

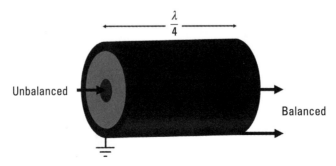

Figure 6.29 Quarter-wave coax balun.

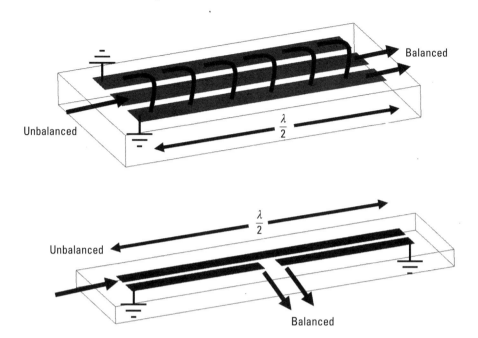

Figure 6.30   Planar balun layouts: edge-coupled (left) and Marchand (right).

transmission line except only one unbalanced end is connected to common ground. The length of the coupled lines is quarter-wave. The structure can also be realized without the second ground trace, but the bandwidth will be reduced. The structure on Figure 6.30's right is a *Marchand balun*. It has similar bandwidth to the edge-coupled design but trades increased size (half-wave versus quarter-wave) for no wire bonds.

When multilayer architectures are an option, the *parallel-plate balun* shown in Figure 6.31 can be used. If the balun is attached to a component that isn't driven from the topside, it offers a convenient end-launch style connection.

Baluns are judged based on the following:

- Amplitude balance: A perfect balun would have equal amplitude signals propagating from each conductor of the balanced port. A real balun will generally have 0.5–1 dB amplitude imbalance.

- Phase balance: A perfect balun would have 180-degree phase shift between the signals propagating from the balanced port. A real balun will generally have ±10-degree phase imbalance.

- Common mode rejection ratio (CMRR): If two identical signals are fed into the balanced port, the 180-degree phase shift between them should cause them to cancel completely. Therefore, in an ideal balun, the

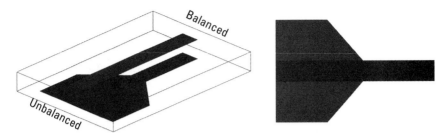

Figure 6.31   Orthogonal view (left) and top view (right) of a parallel-plate balun layout.

CMRR would be infinite. Since real baluns have amplitude and phase imbalance, the CMRR indicates how closely the two signals would cancel. Most baluns have CMRR greater than 25 dB.

There are two additional benefits to using a balun:

1. When combining two equivalent nonlinear components (i.e., a pair of power amplifiers), all even-order harmonics are cancelled. This is due to the out-of-phase nature of the balanced signals. Baluns cancel harmonics over the entire operating frequency range (which can be a decade or more).

2. It is possible to implement an impedance shift as part of a balun. Generally, impedance shifting or impedance matching baluns are expressed by their unbalanced-to-balanced impedance ratio. For example, a 1:4 balun would shift a 50-Ω unbalanced port to a 200-Ω balanced port, and a 4:1 balun would shift a 50-Ω unbalanced port to a 12.5-Ω balanced port. This is useful for matching high-impedance (i.e., antennas) and low-impedance (i.e., large power amplifiers) circuits. Figure 6.32 shows a 1:4 planar balun.

6.10 Mixers

Modern radar would not be possible without mixers. In most systems, the frequencies used to seek and track objects are much higher than analog-to-digital converters (ADC) can process. Therefore, the RF frequency is downconverted to something that can be accurately transformed to digital and processed. Likewise, digital waveforms are transformed to analog using a digital-to-analog converter (DAC) and upconverted to the appropriate RF frequency. This frequency transformation is provided by the mixer. Multiple mixers can be cascaded when large-scale transformation is needed.

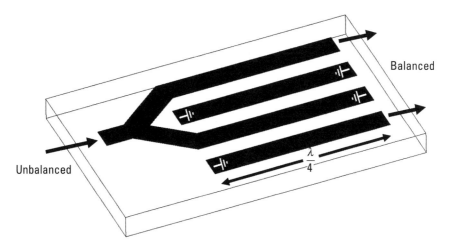

Figure 6.32 A 1:4 planar balun layout.

A mixer is a three-port device as shown in Table 1.2. The three ports are for RF, LO, and intermediate frequency (IF). Mixers are reciprocal devices, so they can be used as both frequency upconverters and downconverters. For down-conversion, RF is the input and IF is the output. For up-conversion, IF is the input and RF is the output (two tones represent high and low side). In both cases, LO (discussed in Chapter 4) is an input that provides the multiplying frequency. The input and output frequencies are shown in Figure 6.33 [8].

For example, 5 GHz at the RF port and 6 GHz at the LO port will produce 1 GHz at the IF port (down-conversion). Alternatively, 1 GHz at the IF port and 6 GHz at the LO port will produce 5 GHz (RF1) and 7 GHz (RF2) at the RF port (up-conversion).

Mixers can be designed in many ways, but two popular designs are the single- and double-balanced mixers shown in Figure 6.34. The hybrid coupler outputs are 180 degrees out of phase, which means they could be replaced with a balun. Double-balanced mixers have the benefit of improved linearity at the

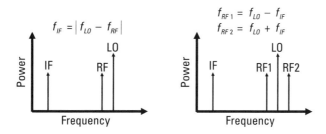

Figure 6.33 Example response of a mixer providing frequency down-conversion (left) and up-conversion (right).

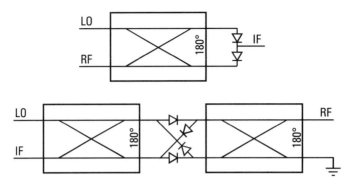

Figure 6.34   Single- (top) and double- (bottom) balanced mixer.

IF port and lower conversion loss (RF power-IF power) than single-balanced mixers [8].

No DC power is needed for a mixer to operate so it is a passive device. However, since it contains nonlinear elements (i.e., diodes or transistors), mixers are also nonlinear. If enough input power is provided, a mixer will saturate just like an amplifier. Additional (usually unwanted) tones will be generated when a mixer becomes nonlinear.

6.11   Antennas

Antennas are components that transform the system impedance (usually 50Ω) to the free-space impedance (377Ω, derived in Chapter 8) to facilitate radiation. Antenna design is a complete and extensive topic in itself and interested readers are encouraged to read [9]. For the purposes of this book, the key definitions are listed as follows.

- Radiator: The part of an antenna that radiates (propagates electromagnetic energy);
- Reflector: An optional part of an antenna that focuses radiated energy in a desired direction;
- Boresight: The direction the antenna is pointing;
- Boresight error (BSE): The difference between boresight and the direction of maximum radiated intensity;
- Beam or lobe: The volume of radiated energy directed away from the antenna;
- Sidelobe: Radiation in an unwanted direction;

- Peak sidelobe ratio: Ratio of the highest sidelobe and the main beam intensity;

- NdB beamwidth: The width of the main beam (in degrees) where the gain is within NdB of maximum (typically, 3 dB);

- Azimuth (AZ): The direction of a reference point looking "left to right" or along the XY-plane (expressed from 0–360 degrees or −180–180 degrees);

- Elevation (EL): The direction of a reference point looking "up and down" or along the Z-axis (expressed from -90° to 90°);

- Range: The distance from the antenna to a reference point;

- Antenna or radiation pattern: A 2D or 3D plot of field strength in space;

- Isotropic antenna: An antenna capable of radiating with equal field strength in all directions with no loss (not feasible in practice);

- Gain: The ratio of the field strength at boresight (typically) compared with the field strength of an isotopic antenna;

- Directivity: The ratio of the field strength at boresight (typically) compared with the field strength of the antenna averaged over all directions;

- Omnidirectional antenna: An antenna capable of radiating with equal field strength in all azimuthal directions (appears as a circle on an antenna pattern plot).

6.12 Current Density Analysis

Compared with active circuits, passive components are relatively straightforward to design. When circuits do not work in simulation as expected, it's often difficult to know where to begin diagnosis. A great starting point is to look at current density plots. Figure 6.35 was generated using ADS; on the left is a meandered delay line, and on the right is the same delay line with a 1-mil gap (break) in the signal line.

Using ADS, a designer can plot the flow of current through the signal path. In the figure above, the line length is greater than a wavelength so a dark spot appears where the signal is zero. When viewed in time-varying mode, the dark spot (and neighboring light spots) traverse the signal path.

Discontinuities usually show up as an interruption to the signal flow. Areas of concentrated current indicate possible sources of radiation and failure. Adding miters or smoothing sharp corners can reduce high current concentration areas. If it is determined that excessive current is being driven into a narrow line, that signal trace should be widened (preferred) or thickened to handling

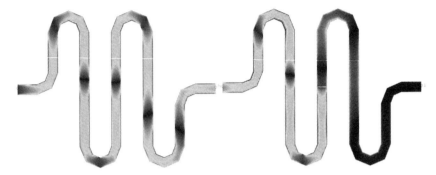

Figure 6.35 Current density plot of perfect meandered delay line (left) and the same structure with a 1-mil gap in the middle.

the current load. Otherwise, the metal trace is likely to heat up, expand, delaminate, and break open.

Current density analysis can also be used to check for unwanted coupling. If two metal traces couple due to their close proximity, signal from one trace will show up on the other. Figure 6.36 shows the pad locations for a simple DC bias network and matching circuit. In the left image of Figure 6.36, coupling is found between the signal line and the DC feed line as indicated by the light grey color on the DC feed. This can be resolved by increasing the spacing between traces, as shown in the right image of Figure 6.36.

Alternatively, the DC feed could be shielded to prevent coupling (discussed in Chapter 8).

Exercises

1. Calculate the optimal right-angle miter for a 50-Ω line on 25-mil-thick substrate of permittivity 6.0.

2. At 8 GHz, what is smaller: a microstrip or lumped-element quadrature coupler? Use 25-mil-thick substrate of permittivity 6.0 for the mi-

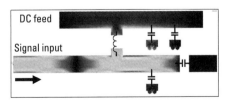
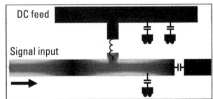

Figure 6.36 Current design plot of signal and bias lines with significant coupling (left) and negligible coupling (right).

crostrip design and 0402 surface-mount components for the lumped-element design. (Hint: Package sizes are presented in Table 3.1.)

3. To protect against open circuit conditions, an attenuator can be added at ports that needs protection. Design a practical 3-dB attenuator.

4. Tones are applied to a Class AB amplifier at 9.5 GHz and 10 GHz. Design a filter using Keysight ADS that can remove the high-frequency second-order products. Use as few elements as possible.

5. An unequal-split Wilkinson power divider is needed with the largest power ratio possible. If the minimum producible signal line width is 5 mils, using 25-mil-thick of permittivity 6.0:

   • What is the largest power ratio that can be achieved?

   • If the substrate could be changed, what qualities should it have to maximize the power ratio?

6. Design bias networks for series and shunt PIN switches at 14 GHz.

7. Use the bias network from the previous exercise to design a 45-degree switched-line phase shifter (include all physical dimensions). Use the same substrate as Exercise 1.

8. Which balun topology is best suited for integrating a pair of surface-mount components to create a push-pull amplifier?

References

[1]   Fano, R. M., "Theoretical Limitations on the Broad-Band Matching of Arbitrary Imped-ances," *Journal of the Franklin Institute,* Vol. 249, January 1950, pp. 57–83.

[2]   Mantaro Product Development Services, Inc., "Impedance Calculators," Internet, http://www.mantaro.com/resources/impedance_calculator.htm.

[3]   Chang, K., I. Bahl, and V. Nair, *RF and Microwave Circuit and Component Design for Wireless Systems,* New York, NY: John Wiley & Sons, 2002.

[4]   Radant MEMS, Inc., "World's First Demonstration of Microelectromechanical Systems -Based X-Band Radar," Internet, http://www.radantmems.com/radantmems/04-06-06.html.

[5]   Kingsley, N., and J. Papapolymerou, "Organic 'Wafer-Scale' packaged miniature 4-bit RF MEMS phase shifter," *IEEE Transactions on Microwave Theory and Techniques,* Vol. 54, No. 3, March 2006, pp.1229–1236.

[6]   Collin, R., *Foundations for Microwave Engineering,* New York, NY: IEEE Press, 2001.

[7]   Pozar, D., *Microwave Engineering,* New York, NY: John Wiley & Sons, 1997.

[8]   Marki, F., and C. Marki, "Mixer Basics Primer," Internet, http://www.markimicrowave.com/assets/appnotes/mixer_basics_primer.pdf.

[9]   Balanis, C., *Antenna Theory: Analysis and Design,* Hoboken, NJ: John Wiley & Sons, Inc., 2005.

Selected Bibliography

Maas, S., *Practical Microwave Circuits,* Norwood, MA: Artech House, 2014.

Wang, G., and B. Pan, *Passive RF Component Technology: Materials, Techniques, and Applications,* Norwood, MA: Artech House, 2012.

Mongia, R., I. Bahl, and P. Bhartia, *RF and Microwave Coupled-Line Circuits,* Norwood, MA: Artech House, 1999.

Part III
Higher-Level Integration

7

Microwave Integrated Circuits

The RF front end is a general term for all components between the antenna and the ADC or DAC. For the transmitter, this includes the frequency upconverter, power amplifier, filter(s), and antenna. For the receiver, this includes the antenna, LNA, filter(s), and frequency downconverter. Both sides leverage the passive circuit elements presented in Chapter 6 to modify the signal(s) as needed.

With an understanding of the individual passive and active circuit elements, the focus of the book turns toward integration. Just like quality ingredients must be combined properly to make a great meal, the best RF components must be integrated intelligently to make a great radar.

This chapter discusses different types of component, subassembly, and packaging form factors. Since radars can operate pulsed or CW, design considerations for both types of operation are included. In addition, the chapter presents techniques for utilizing simulators to quickly and accurately determine module performance and lists the most commonly used radar component specifications, which are often derived from military standards. Finally, the chapter reviews strategies for improving manufacturability and yield, since a radar that outperforms expectations but cannot be manufactured is unacceptable.

7.1 Component Integration

In most cases, a front end is assembled by integrating components from mixed technologies; usually some components are purchased COTS, and some are produced in-house. This section presents the various types of components that are usually encountered and the best ways to integrate them.

7.1.1   MMIC

An RF circuit that is fabricated on a single piece of semiconductor substrate is called a *monolithic microwave integrated circuit* (MMIC). The word monolithic is derived from the Greek words monos (single) and lithos (stone). As discussed throughout the book, common RF semiconductors used in radar include GaAs, GaN, and InP. Figure 7.1 presents a MMIC.

Although there are MMICs that only contain passive circuits, most MMICs are amplifiers of one form or another. Transistors rely on semiconductors to operate, so it makes sense to integrate an amplifier and other functionality with it.

Arguably the most beneficial feature of a MMIC is the precise fabrication tolerance offered and inherent compactness. MMIC metal traces can have submicron tolerance, whereas printed circuit board manufacturers generally charge a premium when less than one mil (submil) tolerance is needed (1 mil =

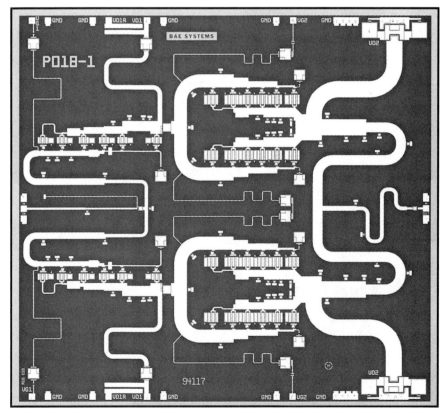

Figure 7.1   Image of a GaAs MMIC. (From Meharry, D. E., et al., "Multi-Watt Wideband MMICs in GaN and GaAs," IEEE/MTT-S International Microwave Symposium, June 2007, pp.631–634. © 2007 IEEE. Reprinted with permission.)

25.4 μm). Semiconductor substrates are nearly flawless, whereas printed circuit boards usually have 10% tolerance on both thickness and permittivity. With this level of precision, a MMIC circuit can be fully optimized to minimize parasitic effects, shrink size, and realize premium performance.

As is usually the case in engineering, there are trade-offs to using MMICs. Foundries that produce MMICs have strict rules on how circuits can be laid out. It takes time to ensure that all design rules are followed and that no unwanted coupling can happen between metal traces. Consequently, the design process for MMICs is typically longer, which adds engineering cost to the effort. A printed circuit board can be fabricated in days, but a MMIC requires at least six weeks to fabricate. Additionally, semiconductors are more costly than printed circuit boards, so this raises the material expense. For high-power applications, dissipating the heat generated from a small area can be a challenge (discussed in Chapter 8).

7.1.2 Hybrid

A hybrid circuit closely integrates two or more components (or subcomponents) to perform a single function. For example, a single diced transistor (called a *discrete transistor*) can be bonded to external circuits that contain the input- and output-matching circuits. Often a hybrid is used instead of a MMIC due to the following factors:

- Availability of parts: If existing in-house or COTS parts are being used with incompatible form factors, their interfaces can be matched using a hybrid.

- Lower cost: For expensive semiconductors, such as GaN-on-silicon carbide, matching circuit components can be placed off-MMIC to reduce the size of the semiconductor and lower cost.

- Broader bandwidth: Techniques that offer broader bandwidth by designing multisection components, such as multistage couplers or matching networks, may be more easily realized off-MMIC.

- Thermal considerations: For high-power applications, combining multiple components of moderate power level spreads the heat better than producing a single high-power component.

- Better manufacturing yield: As MMICs become larger, they become more difficult to fabricate and handle. Keeping the MMIC small increases foundry yield and assembly yield.

- Form factor: Greater flexibility to the overall size and shape is provided by a hybrid so fitting to an existing form factor is much easier.

When integrating multiple components, it is important to ensure that ground path integrity has been maintained. Figure 7.2 shows a side view of two MMICs being integrated from different substrates. The signal path integrity between MMICs is maintained by keeping wire bond lengths short. However, the ground path must travel down through the first substrate, across the gap, and up through the second substrate. For thick substrates, this can add significant phase difference between the signal and ground, which can excite higher-order modes. To avoid this, bond wires can be added topside to connect the grounds from both MMICs. Most MMICs are designed with coplanar waveguide launches to enable direct topside ground connections.

7.1.3 Multichip Modules (MCMs)

Multiple MMICs, packaged components, and/or hybrid circuits can be integrated to create a MCM. In this configuration, multiple functions are performed, such as filtering, built-in test, and amplification. As unpackaged MMICs (also called bare die) are becoming more widely available, COTS MCMs are becoming more popular.

To make assembly, testing, and debugging easier, MCMs are usually the building blocks for higher-level assemblies. Figure 7.3 shows an example of what a MCM could look like. Multiple types of packages are integrated together (discussed in Section 7.1.4) to form a single module. A substrate is added to provide transmission lines to connect the components together (transmission lines not shown). Bare die are connected to other components through wire bonds. (For simplicity, only four are shown.) Each of the dark gray components is a hybrid circuit that can be individually tested for compliance before integrating into the next higher-level assembly.

Components that generate a lot of heat cannot be attached directly to the substrate. From the thermal conductivity of the material, the rise in temperature due to heat generated from the component can be determined (discussed in Chapter 4). If the temperature rise is above the maximum operating

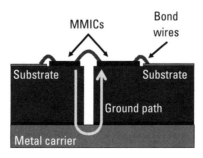

Figure 7.2   Side view of two MMICs connected topside with excessively long ground path.

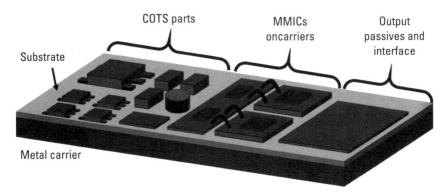

Figure 7.3   An MCM.

temperature, then a more efficient thermal path to the heat sink (in this case the metal carrier) must be implemented.

One method for improving the thermal path to the heat sink is to attach the heat source (i.e., MMIC or hybrid circuit) on top of a via field. A via field, shown in Figure 7.4, is a region of the substrate that has been tightly packed with metal-filled vias. This is generally easy to implement in the substrate and does not add any board-level assembly steps.

For optimal heat removal, the best method is to attach the heat source directly to the heat sink. In this example, that requires removing a section of the substrate and mounting the MMIC or hybrid directly to the metal carrier. This is shown in Figure 7.5. Since the MMIC or hybrid resides inside the substrate, the wire bonds connecting to the external components will be longer by at least the substrate height. This difference must be taken into consideration.

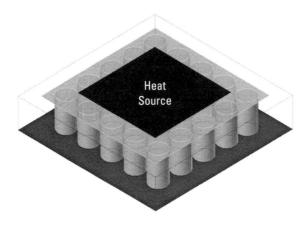

Figure 7.4   Heat source with improved thermal path using a via field.

Direct attach
to carrier

Figure 7.5   Heat source with improved thermal path using a direct carrier attach.

For example, the added inductance from adding 25 mil of wire bond length is sufficient to shift a 10-GHz resonance to 11.8 GHz.

7.1.4   Packaging Options

It is not uncommon for a package to add 0.5–1-dB loss (not to mention a reactive element) to a component's performance. Unfortunately, the effects of packaging are often overlooked or underestimated. As component performance improves with new materials and techniques, packaging effects can drive overall performance. Although high-quality and high-frequency packaging options are available, they can be costly. Some of the more popular packaging options (bottom view) are shown in Figure 7.6 and discussed in this section.

Flange-Mount Package

This is the industry standard for any component that operates with more than 10W DC power (or dissipated power, PDC). The component is mounted directly to a metal carrier, which is then screwed onto the heat sink. Since there is direct metal-to-metal contact, this package offers superb thermal and electrical performance. Flanges extend from the input and output to be soldered to the substrate. A ceramic or plastic lid can be added to encapsulate the component. If the lid is ceramic, the package can be made hermetic (blocks moisture). These packages can be designed to operate well into millimeter-wave frequencies.

The downside to this package is the size. Since it is screwed down, surface area is consumed by the attachment method. Most flange-mount packages have two or four ports, so DC and RF signals share a port.

QFL Package

For components that operate with less than 10 W DC power, the QFL package is a more easily integrated option. As shown in Figure 7.6 (center image), many

Figure 7.6 Bottom view of flange-mount, quad flat lead (QFL), and quad flat no-lead (QFN) packages.

ports are available on a QFL so DC and RF ports can be handled separately. Since the leads extend away from the package, they are easy to solder into and out of modules. Often the leads are formed with a bent or gull wing shape so that the bottom of the leads are flush with the bottom of the package. In the center of the package bottom is a metal pad called a *paddle*, which serves as both the thermal and electrical path to ground. This package can also support millimeter-wave frequencies.

Other than the power-handling limitation, this package is also limited by the number of ports it can handle. Since all leads are placed on the perimeter, ports must be spaced adequately to prevent coupling.

QFN Package

If the leads that extend beyond the edge of a QFL are trimmed away, the resulting package would be a QFN package. The footprint size for this package is smaller than a QFL, but it is more difficult to assemble into a module. Some QFN packages have castellated ports, which means the metal wraps around the edge. This makes assembly easier.

This package has the same power and port limitations as the QFL. However, additional ports can be added by changing the interface to a ball grid array (BGA), which uses a series of bumps on the backside of the package as ports. Since the distance from the component to the port is minimized, a BGA offers low port inductance and higher operating frequencies.

Metal Package

A metal package (not shown in Figure 7.6) can be made to fit any size component. If hermeticity is not an issue, an inexpensive way to protect a component is to screw a folded-metal lid over it. Stainless steel is often used because it is easy to machine. Alternatively, a hermetic package can be implemented by welding the metal lid to a metal carrier.

Small metal covers are often called *dog houses* because of their resemblance to the namesake. They provide a convenient mechanism for electromagnetic shielding (discussed in Chapter 8).

Plastic Package

Similarly to a metal package, a plastic package can be extruded to fit any size component. Due to their light weight and low cost in volume, they are becoming increasingly popular. Although the lid material is plastic, the inside can be metal-coated to become hermetic. Additionally, the plastic can be formulated to prevent static or provide magnetic or electric shielding (discussed in Chapter 8). Some plastic packages have even been formulated to pass U.S. military standards (discussed in Section 7.3).

7.2   Packaging Model

A simple model for a symmetric package is shown in Figure 7.7, where the subscript "Ext" is for external parasitics, "Int" is for internal parasitics, and "P" is for generic parasitics.

These values are generally determined empirically by replacing the component with a 50-Ω thru line and fitting the parameters to measured S-parameters. At low frequency, the inductor and resistor values can be determined since the capacitors are essentially open circuits. Then, the capacitors can be tuned to meet the higher-frequency behavior.

To test the model, the component can be replaced by a short to ground. The modeled and measured S11 magnitude and phase should match. If needed, the simple model can be expanded to include delay lines (to match phase) and a single input-to-output capacitance (if port-to-port isolation is not sufficiently high to be negligible).

7.3   Designing for U.S. Military Standards

The United States has created a library of documents that clearly define how military systems must be built and tested. They cover every detail from how

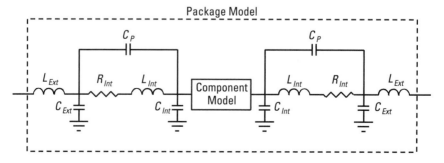

Figure 7.7   Simple model for a symmetric microwave package.

long silver epoxy should be cured through how resilient a system must be to fungal growth. For components, there are several commonly referenced military standards for components. (Note that when searching for these documents, letters are often added after the name to denote the version.)

MIL-STD-883 entitled "Test Method Standard—Microcircuits" [1]

Scope: This standard establishes uniform methods, controls, and procedures for testing microelectronic devices suitable for use within military and aerospace electronic systems including basic environmental tests to determine resistance to deleterious effects of natural elements and conditions surrounding military and space operations; mechanical and electrical tests; workmanship and training procedures; and such other controls and constraints as have been deemed necessary to ensure a uniform level of quality and reliability suitable to the intended applications of those devices.

MIL-PRF-19500 entitled "General Specification for Semiconductor Devices" [2]

Scope: This specification establishes the general performance requirements for semiconductor devices. Product assurance is provided by effective screening, conformance inspection, and process controls to mitigate risk.

MIL-PRF-38534 entitled "General Specification for Hybrid Microcircuits" [3]

Scope: This specification establishes the general performance requirements for hybrid microcircuits, MCMs and similar devices, and the verification requirements for ensuring that these devices meet the applicable performance requirements.

MIL-PRF-38535 entitled "General Specification for Integrated Circuits (Microcircuits) Manufacturing" [4]

Scope: This specification establishes the general performance requirements for integrated circuits or microcircuits and the quality and reliability assurance requirements, which are to be met for their acquisition. The intent of this specification is to allow the device manufacturer the flexibility to implement best commercial practices to the maximum extent possible while still providing product that meets military performance needs.

MIL-STD-1835 entitled "Electronic Component Case Outlines" [5]

Scope: This standard establishes and maintains a compilation of electronic component case outlines and should be useful to all levels of manufacturing that culminate in the production of reliable and logistically supportable electronic equipment. The purpose of this standard is to assure complete mechanical interchangeability of all electronic component case outlines of a particular style and type, regardless of source, commensurate with the requirements of high-density electronic equipment manufacturing.

MIL-STD-461 entitled "Requirements for the Control of Electromagnetic
Interference Characteristics of Subsystems and Equipment" [6]

Scope: This standard establishes interface and associated verification requirements for the control of the electromagnetic interference (emission and sus-
ceptibility) characteristics of electronic, electrical, and electromechanical equipment and subsystems designed or procured for use by activities and agencies
of the Department of Defense. Such equipment and subsystems may be used
independently or as an integral part of other subsystems or systems.

System designers tend to specify which items within the standard apply
to a particular system. Testing for military standard compliance is expensive, so
it is best to avoid testing irrelevant standards (i.e., testing operational performance at high altitude on a ground vehicle platform). The military standards
we describe here total 1,734 pages of detail, so this section will discuss only the
key design parameters.

7.3.1 Robustness

Military radar must handle extreme environmental, mechanical, and electrical
stress (which is why there are so many lengthy military standards). A certain
level of robustness must be met, and a component is only as strong as the weak-
est area. It is important to differentiate between the following:

- Safe operating limit: The stress level at which the component can operate over the expected lifetime without degrading performance;

- Deteriorated operating limit: The stress level at which the component
  can operate over the expected lifetime with degraded performance;

- Survivability limit: The stress level at which the component can operate
  but over which any additional stress will cause permanent damage.

These limits determine the subcomponent ratings. For example, all capacitors have a maximum safe operating voltage and a survivability voltage limit.
Suppose that a capacitor can operate for 1×10^6 hours at 100V and has a breakdown (survivability) voltage of 150V. If analysis shows that under all operating
conditions and assembly tolerances (see the Monte Carlo analysis in Chapter
8) the peak voltage at a particular node is 98V, then that capacitor is properly
rated for use at that node. In most cases, designers use a safely factor of two to
three times rated value to provide margin. So, in this case, it would be better to
choose a capacitor with safe operating voltage of 200–300V.

Practical Note

Be wary of excessive derating. If a design engineer determines that a capacitor at a particular node will never experience voltages above 50V, there is no need to choose a component rated for 500V. Larger-rated parts are usually physically larger and have more parasitics than lower-rated parts. Manufacturers also derate their components so there is no need to over derate.

Substrates also have a rated value that is sometimes overlooked. The dielectric strength (or electrical strength) is the voltage that a material can withstand before dielectric breakdown occurs. It is the same principal that causes high voltages to arc through air. The likelihood of voltage breakdown can be aggravated by surface moisture and contamination. Once breakdown occurs, the substrate is degraded permanently. The dielectric strength of a substrate (ζ_{max}) can be calculated by:

$$\zeta_{max} = \left(\frac{V}{d}\right)_{max} \tag{7.1}$$

where V is the voltage (V) and d is the substrate thickness (m). Dielectric strength values for microwave materials typically range between 20 and 30 kV/mm (500–750 V/mil).

When attaching a MMIC, hybrid, or substrate to a carrier in a military environment, there are four principal options:

- Mechanical: Attaching two materials through purely mechanical means (i.e., bolting them together) is rarely a good idea. Surfaces are never perfectly flat so air pockets form and prevent good thermal and electrical contact. The problem is exacerbated as temperature change causes expansion and contraction. Any sort of sustained vibration will eventually wear down the components. Mechanical attach should never be used at RF frequencies or higher.

- Film: Sheet epoxy, sheet solder, and indium foil are popular due to their ease of use. They are comprised of thin sheets of material that can be cut to size, placed between the components being attached, compressed, and heated until molten. Once the heat is removed, the film solidifies, and attachment is made. Although convenient to handle, films tend to have low thermal and electrical conductivity compared to paste epoxy or solder, which can cause a rise in operating temperature and higher loss. Indium foil does not require heating for attachment, but it does suffer from mechanical creep (the condition where repeated stress causes degradation).

- Epoxy: Silver-loaded epoxies are pastes that can be spread by hand or with an automated system to the areas needed. Since they are silver-loaded, thermal and electrical conductivities are high. They are also rich with lead, which makes them soft and flexible. This is advantageous for applications with large temperature swings or vibration requirements. Some epoxies come as two-part chemicals that must be precisely mixed right before application. They must also be stored cold and have a short shelf life. At high temperatures, they can become brittle and break down electrically so they are not applicable for high-power applications (as determined by the manufacturer's maximum operating temperature).

- Solder: Au/Sn (80% gold/20% tin) solder is the staple for military applications. It is robust and offers excellent thermal and electrical performance. It works well in high-power applications. It is usually applied in a bell jar with nitrogen- or hydrogen-forming gas. The consistency is not as smooth as epoxy, so it can be challenging to apply. It is also harder than lead-based solders, so it has less mechanical flexibility under extreme conditions.

7.3.2 Operating Stability

Just as a new pair of shoes needs to be "broken in" before they wear well, most semiconductors need to be exercised before deployment. This process, called *stabilization* or *burn in*, is discussed in Chapter 9. Essentially, the conducting junction or channel matures as current passes through it. Once a stable path is formed, the electrical performance will not change for the duration of the operating lifetime provided the conditions are unchanged (i.e., same or lower bias, RF drive level, or temperature).

Unfortunately, many system specifications require 100-hour stabilization at 100°C (legacy numbers from older military standards). For a GaAs component, this generic process can be so severe that it degrades performance and consumes usable lifetime. Similarly for a GaN component, this process may drive up the cost of testing and have an insignificant effect (higher temperature may be needed). It is important to use military standards judiciously.

7.3.3 Environmental Considerations

Radars can be found everywhere from deep under water to deep space, so the range of possible environmental conditions is tremendously large. Systems (and therefore components) must operate regardless of the environment. Some of the more problematic environmental conditions are addressed as follows.

Temperature

The typical operating temperature range for a military system is –40 to +85°C, although some span –55 to +125°C. Since most electrical parameters degrade as temperature increases, performance is usually determined by the highest operating temperature. Electrical stability (K_{factor} from 4.20) is more difficult to achieve at low temperature, so analysis must be performed at the lowest temperature. If cold models are not available, a simple analysis can be performed by adding a gain stage equivalent to 0.016 dB/°C per stage to account for the added gain at low temperature. For example, a two-stage amplifier will add approximately [0.016 dB/°C · (85 – –40°C) · 2 stages =] a 4-dB gain between hot and cold. It should be verified that the active components are still stable with the added gain (Section 4.2.1).

All materials should be checked to ensure that the operating temperature range is within its safe operating range. Keep in mind that active components dissipate heat, so the temperature a component experiences could be higher than the external temperature. Some materials, like polymers, become brittle at low temperature. Others, like solder, can reflow (become liquid) at high temperature. There are also materials, like indium, that degrade with temperature cycling (a behavior called *creep*). Fortunately, material blends are usually available that provide extended temperature range.

An assortment of tests can be performed to evaluate temperature effects on a component, including the following:

- Thermal shock: A sample is cycled between submerging in hot liquid and submerging in cold liquid. The sudden change in temperature accelerates the effects of thermal mismatch between materials.

- Thermal cycling: A sample is cycles between hot and cold air with a soak time (nominally 30 minutes) after each step. The gradual change in temperature accelerates the effects of thermal creep in the component.

Moisture

An enclosure that is hermetic is impervious to gas and is formed by seam-sealing or laser-welding two metals, ceramics, and/or glasses together. If gas particles cannot enter (or escape) then neither can moisture or particulates. The military (and industry) standard leak test requires filling a cavity with helium and then measuring for leaks. Helium is used because it is an inert gas and the second smallest element on Earth (after hydrogen). This is called a *fine leak test*, and MIL STD 883 specifies the amount of allowable leakage. Alternatively, a gross leak test requires dipping the sample in a clear liquid and watching for bubbles.

> **Practical Note**
>
> Most MMICs are covered by a protective insulating layer called a *passivation layer*. It is usually made from a thin layer (100–300 nm) of silicon nitride. Contrary to popular belief, MMICs that are properly passivated can operate with moisture on the surface. However, due to water's dielectric constant of approximately 80 (it varies greatly with frequency), surface moisture can dramatically change the electrical performance. That is the real reason MMICs are usually enclosed in a hermetic package.

Surface moisture can also change the performance of substrates. As moisture is absorbed into the material, the substrate is being loaded with a high-dielectric material (again, approximately 80). This raises the effective dielectric of the material, which in turn changes the impedance of the transmission line. Fortunately, the effective dielectric constant is not the weighted average of the two individual permittivities. For example, a 25-mil-thick substrate that is half permittivity 2 and half permittivity 80 is not permittivity 41. It is actually approximately 4.2 (determined through full-wave analysis). This may seem counterintuitive, but it is similar to combining 2-Ω and 80-Ω resistors in parallel. The result is approximately 1.95Ω; the effect of the 80-Ω resistor is almost negligible.

Some materials, like liquid crystal polymer, are hydrophobic and do not absorb moisture. Instead, moisture passes right through them like water through a sieve. These materials maintain electrical performance in high-moisture environments.

Corrosion

Other than inert gasses, nearly every other substance is corrosive to electronics. Hydrogen causes damage to pHEMT devices. Salt mist and oxygen can break down metals. Even fine particulates can cause excessive scratching over time. Just as MMICs are coated in a protective passivation layer, other components can be coated in protective layers as well. Parylene and all of its variations are a popular option for radar components because of their low dielectric constant (< 3). It can be added with little thought as to the electrical ramifications.

Parylene can be easily deposited by vapor deposition or dipping in liquid so oddly shaped objects are not difficult to coat. It offers a corrosion barrier against strong acids and bases and can handle high temperature (> 300°C).

7.3.4 Electrical Considerations

We have already considered performance changes due to dielectric loading from moisture. Additionally, environmental effects can change the performance of the components connected to the input and output, which can alter the matching condition. Generally this is only an issue for active components. Instead of providing a fixed system impedance (typically 50Ω), the component may be terminated with a changing impedance within an N:1 VSWR circle. For example, a component may be required to operate within a 3:1 VSWR circle (return loss can be as bad as 6 dB over all phase angles). Some protection can be added by incorporating a small attenuator at the port, but this adds loss. Alternatively, a hybrid coupler can be incorporated as discussed in Section 4.3.2.

Components may need to be protected against DC or RF power spikes. This can be caused by system failure or electronic countermeasures targeting the radar. DC spikes can usually be mitigated by designing smoothing circuits into the DC supply or power-conditioning circuit. Short DC bursts are more difficult to handle, because the transient response (rise and fall time of the spike) can be similar to the RF operating frequency. Filtering the response to remove spikes is therefore not possible. Instead, high-breakdown materials, like GaN, can be used for added robustness. GaN operates at 28V or 50V, but the breakdown voltage is usually 150V or higher. Even when the peak voltage is on the drain, there is still ample margin in the event of a spike.

In applications where high-breakdown materials cannot be used, oversizing the device will increase the potential power handling. This will come at the expense of performance since the amplifier will not operate near saturated power.

7.3.5 Mechanical Considerations

Whether mounted on a warfighter, military vehicle, aircraft, or satellite, systems and components are going to be exposed to vibrations, acceleration, and various forms of drop shock. For the most part, microelectronic circuits (and subcircuits) are so small and lightweight that these effects are negligible. Where it becomes an issue is at the system level, where many large components are being combined together and can flex. Also, the physical size can be sufficiently large to have a mechanical resonance. If the vibration frequency matches the resonant frequency, excessive wear or fracturing can occur. This can be avoided by performing a full structural analysis and testing for those physical conditions.

7.4 Designing for Pulsed Radar

Radars operate in either continuous wave (CW) or pulsed mode. This section discusses the impact of pulsed operation on component design.

7.4.1 Radar Terminology

In pulsed mode, a radar toggles between transmit and receive functionality as shown in Figure 7.8.

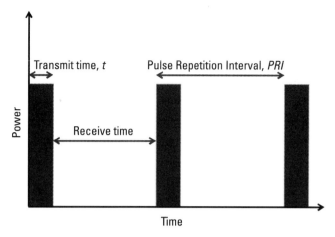

Figure 7.8   Radar transmit and receive behavior in pulsed mode.

The duration of the transmit cycle is also called the *pulse width* (τ). The periodicity of transmit pulses is called the *pulse repetition interval* (PRI). The reciprocal of *PRI* is the pulse repetition frequency (PRF). The duty cycle is the fraction of time the radar is transmitting and is calculated by:

$$D = \frac{\tau}{PRI} = \tau \cdot PRF \tag{7.2}$$

where D is the duty cycle (percentage), τ is the pulse width (s), *PRI* is the pulse repetition interval (s), and *PRF* is the pulse repetition frequency (Hz).

PRF is a very important factor in designing pulsed radar. If the PRF is too high (meaning that pulses occur too often), then the echo from one pulse reflecting from a far target may not be distinguishable from the echo coming from a later pulse on a closer target. This is known as *range ambiguity*. Similarly, if the *PRF* is too low the Nyquist condition for Doppler may not be met. This is known as *Doppler frequency ambiguity* [7]. Radar designers must find the right balance.

The average power for a pulsed radar can be calculated from:

$$P_{avg} = P_t \cdot D = P_t \cdot \frac{\tau}{PRI} = P_t \cdot \tau \cdot PRF \tag{7.3}$$

where P_{avg} is the average power level (W) and P_t is the transmit power level (W).

From (7.3), it can be seen that to achieve the same level of average power, pulsed radar must use much higher peak power level.

In modern radar systems, the transmit bandwidth of the radar signal can be greater than the reciprocal of the pulse width when utilizing spread spectrum techniques (referred to as *pulse compression* in radar parlance).

7.4.2 Component Design

Fortunately for components, operating in pulsed mode reduces the average DC and RF power level, which lowers the operating temperature. In fact, applications with very narrow pulse width and low duty cycle can completely mitigate the effects of temperature (known as an *isothermal condition*).

Practical Note

It is common to see a 0.5–1-dB increase in output power and a 1–2-dB increase in gain when an amplifier changes from CW to pulsed mode. Efficiency also increases as a result of the increased gain.

To operate in pulsed mode, there are three design factors to consider.

- Time and phase alignment: When combining signals in CW mode, phase alignment is important to minimize combining loss. In pulsed mode, if two signals are not time-aligned, they could miss each other entirely. This is particularly important for narrow pulses. Full-wave simulation (discussed in Section 7.5) can help ensure good time alignment.

- Stability: Amplifiers that are unstable at CW may be stable when pulsed. If the oscillation is caused by a feedback loop (discussed in Chapter 8), the pulse width could be shorter than the time required for the feedback to occur. Even if a component is intended to only operate pulsed, it is a good idea to verify CW stability, even if only using S-parameters. It is common to debug circuits using CW signals (even if only at low power) to make testing easier.

- RF blocking (choking): If the drain or gate voltage is being pulsed to conserve prime power, this can become an issue for the RF-blocking circuitry. In CW mode, the RF choke uses a series inductance and shunt capacitance to ground to pass DC and block RF. Inductors treat fast DC pulses the same way they do AC signals. This is shown in Figure 7.9. As the series inductance increases, the voltage pulse takes longer to turn on and off. This could degrade or potentially cripple a pulsed radar. A balance must be found that provides enough isolation between the component and the pulsed voltage source without preventing the pulse from propagating through it.

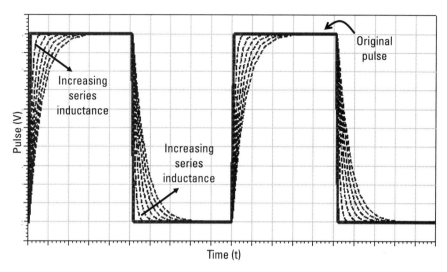

Figure 7.9 Degradation of pulse as series inductance from RF choke increases.

Whenever components are not operating (say, the LNA during transmit mode or the PA during receive mode), they are often biased down by pinching off the gate or applying 0V to the drain. This is known as *blanking*, and it protects the components and can reduce prime power. Component specifications will often include a blanking time, which indicates how quickly the component must turn on and/or off when triggered. Blanking is especially beneficial to class A amplifiers since they consume prime power even when no RF is applied.

7.5   Taking Advantage of Simulators

It's hard to imagine how complex designs were ever done before the days of computers and simulators. Today, realizing "first-pass success" isn't just a marketing term; it is achievable when a simulator is used by someone who knows how to apply it properly.

In general, simulators can do the following:

- Determine small and large-signal performance (provided that suitable models are available);
- Optimize component values;
- Show how varying one (or more) parameter can affect performance;
- Perform sensitivity analysis (discussed in Chapter 8);
- Analyze components in multiphysics environments (i.e., electrical, thermal, and mechanical);

• Generate initial designs based on traditional approaches.

In general, simulators cannot do the following:

• Invent new architectures;
• Trade off multiple approaches and materials;
• Find ways to make a design more compact;
• Determine what is manufacturable;
• Investigate cost-saving options.

A widely used RF simulator is Keysight ADS [8]. Simulation examples shown throughout this book and the remainder of this section were generated using ADS.

7.5.1 Passives

One nice feature of ADS is that it offers built-in designs that it calls Design-Guides. Nearly every passive component imaginable has a dedicated Design-Guide for both the lumped and distributed form. A designer can select the component needed from the drop-down menu, enter the key specifications (i.e., frequency, bandwidth, and ripple), and in seconds the design is done. The design may not meet all specifications, but it can almost always provide something close. Then, the designer can make tweaks and optimize from that starting point. The distributed and lumped-element band pass filters shown in Chapter 6 were designed using the ADS Filter DesignGuide.

In addition to DesignGuides, ADS also offers a utility called LineCalc that performs a similar function, but for more standard designs. As an example, an 8-finger 3-dB Lange coupler was designed using LineCalc that operates at X-band. Lange couplers are notoriously tricky to design, but ADS was able to generate the design shown in Figure 7.10 instantly.

Without changing the design, the layout in Figure 7.10 was simulated, and the results are shown in Figure 7.11. Return loss (S_{11}) and isolation (S_{41}) are better than 29 dB across X-band. The thru (S_{21}) and coupled (S_{31}) loss are within ±0.3 dB of each other.

Other than components, ADS can also be used to generate matching networks. Any two impedances can be entered, and ADS will provide a list of possible matching networks. It is certainly much faster than using a Smith chart.

7.5.2 Actives

Chapters 4 and 5 show that amplifier design is rooted in circles (i.e., power, gain, efficiency, stability, noise, and match). Fortunately, ADS can generate

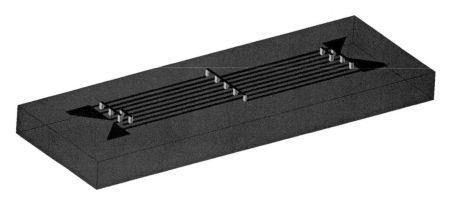

Figure 7.10   Autogenerated Lange coupler from ADS.

those circles from a device model. There are even built-in DesignGuides that generate those circles automatically.

ADS has several useful features for analyzing stability. For looking at component-level stability, ADS will calculate the four stability parameters discussed in Section 4.2.1 (K_{factor}, $K_{measure}$, μ_{load}, and μ_{source}). Additionally, ADS offers a two-port stability probe (*SProbe*) that can be placed on any node in a circuit to determine if it is unstable at that node. The probe is most accurate when placed directly on the gate and drain nodes. Sometimes losses between the transistor and the input/output ports can mask an instability. Adding an SProbe directly on the transistor prevents this from happening.

7.5.3   Full-Electromagnetic (EM) Simulation

Schematic simulators operate by building a block diagram of lumped elements, distributed elements, and other components. Figure 7.12 presents a schematic for a two-way Wilkinson divider. Each block has a behavioral model that the simulator uses to calculate the integrated performance. The "S-parameters" block lists the frequency range to simulate over. The "MSub" block includes information about the substrate (in this case 25-mil-thick alumina). Any coupling that may occur between TL2 and TL3 would not be captured in this simulation.

To improve accuracy, the substrate components (i.e., the MLIN transmission lines) can be simulated using full EM analysis (also called *full-wave analysis*). Full-EM simulators calculate the electric and magnetic fields determined by Maxwell's equations [(1.8), (1.11), (1.13), and (1.16)] including all time-varying and frequency-dependent behaviors. Coupling between structures would be accurately captured. Keysight offers a design package called FEM Simulator that allows a user to draw a circuit in three dimensions, simulate the

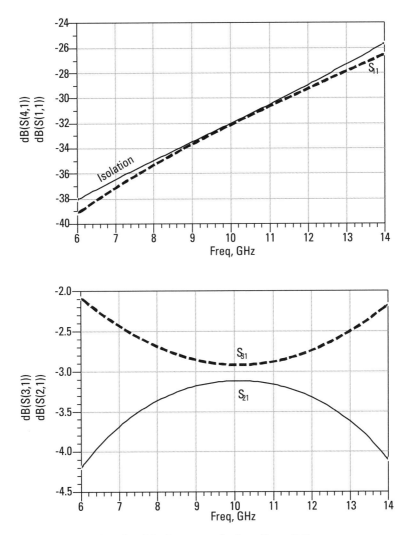

Figure 7.11 Simulated results of the Lange coupler from Figure 7.10.

circuit, and view the waves propagating through the structure. For complicated, three-dimensional circuits this is an incredibly powerful tool.

For designers who need more fidelity than can be achieved with a behavioral model, but who do not need something as sophisticated as a three-dimensional model, there is an intermediate option. These simulators are called *2.5D, quasi-3D,* or *planar 3D.* They run faster than full-EM simulators, but not as fast as behavioral simulators. For the circuit in Figure 7.12, a 2.5D simulator is ideal. Substrate losses and coupling between structures would be captured. A designer could watch the RF current flow through the transmission lines (if needed) to diagnose issues.

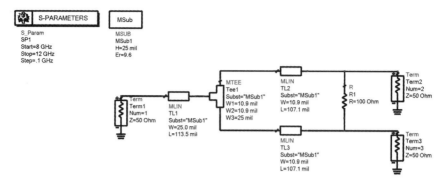

Figure 7.12 ADS circuit simulation setup for Wilkinson power divider.

7.5.4 Manufacturing Assessment

If a component could be built exactly as modeled, then it would work as expected every time. Unfortunately, this is never the case. Even the best manufacturers have tolerances that must be accounted for. Substrates typically vary in thickness and permittivity by 10%. Milled mechanical parts are usually within a mil (25.4 μm) of specification. Surface-mount elements (i.e., resistors and capacitors) are rated between 1 and 20%, depending on price. The list goes on and on.

Every component has many elements that can vary, and the best way to simulate a wide range of possible outcomes is with a Monte Carlo analysis. This design step is discussed fully in Chapter 8, but a list of tolerances to consider is presented as follows.

- Bond wire placement: Bonds will not always be placed in the middle of a bond pad, which can change inductance;

- Bond wire height: Bond wire height can be difficult to control and height variations will change the bond wire length and equivalent inductance;

- Gaps due to cutting tolerance: In hybrid circuits, gaps will occur between substrates to allow for manufacturing tolerances. These gaps should be kept small to maintain a good ground connection, but they will vary;

- MMIC process variation: Foundries can usually provide Monte Carlo parameters based on data they have taken on the process variation;

- Line width: Line width variation will change the line impedance;

- Surface-mount element tolerances: Manufacturers of products such as COTS resistors and capacitors should provide a tolerance that varies with price;

- Substrate thickness: Substrate thickness variation will change the impedance and electrical length of the transmission lines;

- Substrate permittivity: Substrate permittivity will change the impedance and electrical length of the transmission lines;

- Port VSWR mismatch: The ports may not always maintain the system impedance (usually 50Ω), leading to performance degradation and instability;

- Temperature variation: The thermal model will have some tolerance, which can change performance;

- Bias voltage: Depending on the power supply, bias voltage can drift with time, which can change performance.

After a Monte Carlo analysis is completed, the range of potential component performance will be determined. From this, a projected manufacturing yield can also be determined (if 90% of all Monte Carlo cases meet the specification, then that is the manufacturing yield—assuming that all possible tolerances have been captured). It can also be determined whether any scenarios can lead to device instability.

Even if a design has an ample performance margin, it doesn't take much unexpected parasitic inductance to shift the operating frequency out of specification. Using a simulator like ADS, a designer can perform a Monte Carlo analysis and if the variation (or yield) is not as expected, designers can determine the main contributors to the variation and possibly modify them to make them less sensitive.

7.6 Manufacturing Practices

Engineering companies are placing a much stronger emphasis on designing for production. This is a design practice where engineers are aware of manufacturing and test limitations during the design phase. Designing with production in mind leads to faster time to market, lower production costs, and ultimately, more competitive bidding. With the radar market becoming so competitive, it is important to address manufacturing practices and how to implement good design strategies.

For efforts that are genuinely research-oriented (bleeding edge), designing for production can stifle creativity. These endeavors are usually several iterations away from production, and only a handful of prototypes will be made in the course of the project to prove feasibility. Parts can usually be handmade or the best parts can be cherry-picked from those available. In these cases, designing for best results rather than production is ideal. In all other cases, including early stage, technology maturation, or alpha/beta phase efforts, designing blind to the limitations of production would be a mistake.

7.6.1 Manufacturing Essentials

The words manufacturing and production are interchangeable. Manufacturing is derived from the Latin words manu factus meaning made by hand. A range between 70 and 80% of the cost of product development and manufacturing is determined during the initial design phase [9]. Engineering goals should not be restricted to only achieving the best possible performance; they should also include goals to make products that are manufacturable (i.e., high-yield and low-cost). There are several things to consider:

- Performance requirements: If the specifications are so aggressive that only the top-performing components will be used, this will drive up cost;

- Manufacturing method: Low-volume components or anything built by hand will have a different manufacturing tolerance than high-volume or automated manufacturing;

- Operating lifetime: Some military platforms are designed to operate for decades (the B-52 bomber is famous for having a 60+ year service record and counting). Components must be selected that can survive under operating conditions for the life of the product;

- Frequency of routine maintenance: Sometimes components cannot be designed to survive under normal operating conditions over the expected lifetime. Normal "wear and tear" may be unavoidable, but compartmentalizing parts that need replacing will improve serviceability;

- Consequences of in-service failure: It's an unfortunate truth that when military systems fail, warfighters may be placed in harm's way. This must be considered when determining the margin that a component has over its specification;

- End-of-life disposal: An often overlooked item, choosing recyclable materials, is becoming more important as environmental awareness increases;

- Cost and price: The cost is the financial requirement to produce the component, whereas the price is what the components sell for. Profit is equal to price minus cost. Since price is often market-driven, to increase profit, cost must be reduced by controlling manufacturing expenses.

These manufacturing considerations require engineers to think about the following:

• Durability: Ability to withstand wear, pressure, and damage;

• Reliability: Consistent behavior for the operational lifetime;

• Longevity: Ability to survive for a long period of time;

• Cost: Expense associated with raw materials and assembly;

• Environmental impact: How the creation, operation, and disposal effects the environment.

Practical Note

The last example (environmental impact) is often neglected, but it shouldn't be. In 2010, the United States discarded 3.41 million tons of aluminum. At 84¢ per pound, that's $5.7 billion dollars in waste (or $18.52 per American). Fortunately, nearly 20% of that aluminum was recovered [10]. Designing with recyclability in mind should be part of the design process.

7.6.2 Engineering Practices for High Yield

The life cycle of a product includes concept, design, development, production, distribution, use, and disposal. Right from the beginning (the concept and design phases), the entire lifecycle should be considered. For example, changing materials late in the design cycle to lower cost or improve manufacturing may not be possible, since the new materials will need to be reanalyzed for mechanical, thermal, and electrical changes.

Some tips for achieving high manufacturing yield are listed as follows.

• Buy from reputable suppliers and question the source of their raw materials. Trace amounts of lead, silicone, or other impurities can cause failures during operation.

• Avoid ambiguity by placing alignment marks (such as triangles) to indicate how a part should be centered. Although the placement of a MMIC, surface-mount element, or bond wire may be obvious to engineers, these factors are not always clear to manufacturing technicians.

• When possible, keep feature sizes large. For example, a 50-Ω line on 10-mil substrate with permittivity 10 is 9.3-mil-wide. On 25-mil-thick substrate, the line width is 24 mils. If both substrate thicknesses are equally suitable for the application, use the thicker substrate since the wider line width is easier to manufacture.

- Be mindful of electrical and mechanical tolerances and use a Monte Carlo analysis to determine areas of sensitivity (discussed in Chapter 8).

- Whenever possible, use thicker and higher-permittivity material, since thinner and lower permittivity materials are more sensitive to substrate variation.

- Screen substrates to reduce component-to-component variation, especially when quantities are high. Most substrate materials have a ±10% tolerance on both thickness and permittivity. Upon request (and sometimes for a nominal fee), this tolerance can be tightened.

- Individually select or bin materials and elements based on desired performance. For example, use only components within 1% of nominal (referred to as the *family group*) or with peak performance. When combining multiple parts (i.e., two amplifiers power-combined), top-performing components can be paired with bottom-performing components to achieve average performance.

- Having optional tuning structures on-MMIC or on-board; when units fall short on performance, these tuning structures can help resurrect a component from the scrap bin. Figure 7.13 shows four common tuning structures: The first image is of a transmission line or DC bias line with grounded traces on either side to support the addition of surface-mount capacitors for tuning or decoupling. The second image is of an open stub with pads that can be connected with wire bonds to extend the electrical length. The third image shows a tuning element called a *trombone* (due to its resemblance of a trombone slide) that can be wired bonded to extend the phase length. The inner connection must also be scratched off. The final image is of a pad field that can be wire-bonded to implement whatever electrical length is needed. In Figure 7.13, maximum phase length has been achieved by wire bonding every pad. Shorter phase delays can be implemented by bonding fewer pads.

Figure 7.13 Common tuning structures: grounded pad, bondable stub pads, trombone line, and pad field.

7.6.3 Designing for MMIC-Level Cost Reduction

If a component includes a MMIC, the MMIC is usually the most expensive part of the component. Foundries are inherently expensive to run and maintain. The material expense is also high, particularly when silicon carbide wafers are used (as is usually the case for GaN).

Unfortunately, the only method a designer has to reduce the MMIC cost is to make the size as small as possible. Since the foundry price is fixed, the more MMICs that can fit on a wafer, the lower the per-component cost will be. Designing for compact size is just a matter of making components small and as tightly packed as possible without compromising isolation.

Unlike modules that can be reworked when there are material or assembly issues, once a MMIC completes fabrication it cannot be modified. From fabrication to product shipment, there are a number of stages where MMIC failure affects overall product yield; they are described as follows.

- Wafer yield (Y_w): Percentage of wafers that complete processing (sometimes wafers are broken during processing due to mishandling or equipment failure).

- Process visual yield (Y_{pv}): Percentage of MMICs that pass visual inspection. [Sometimes foreign object debris (FOD) is introduced during fabrication. Since FOD is localized, fewer MMICs are affected when MMIC size is small.]

- Process electrical yield (Y_{pe}): Percentage of MMICs that pass electrical screening [usually defined by structures called process control monitors (PCMs) that must perform within a tightly controlled window]).

- Screening yield (Y_s): Percentage of MMICs within an acceptable performance window (parts that do not perform "in family" may be suitable for engineering samples but cannot be used in production).

- Assembly yield (Y_a): Percentage of MMICs that are successfully incorporated into the next higher-level assembly. (Sometimes components are damaged during assembly, especially while a process is being matured.)

- Burn-in yield (Y_{bi}): Percentage of MMICs that meet performance requirements after burn-in. (As discussed in Chapter 9, some components will not stabilize and therefore cannot be used.)

The final product yield (Y) is equivalent to the product of these quantities:

$$Y = Y_w \cdot Y_{pv} \cdot Y_{pe} \cdot Y_s \cdot Y_a \cdot Y_{bi} \tag{7.4}$$

Yield values vary greatly depending on the MMIC size, foundry, process maturity, and manufacturer. In general, wafer yield, process visual yield, and process electrical yield usually total 85–95% yield. After screening yield, this drops to 80–90%. Assembly and burn-in yield can drop overall yield to 50% at first, but once the process is matured (especially when volumes are high), yield usually rebounds to 75–90%.

7.6.4   Designing for Module-Level Cost Reduction

The need for cost reduction is self-evident, but how to do it effectively is not. Some tips for designing for cost reduction follow:

- Don't overdesign: A material and design strategy that greatly exceeds what is needed, it is also likely to be unnecessarily expensive. Seek out an alternative.

- Choose materials that are standard products, industry standard, or produced in large quantities: Competition and volume will drive down the price

- Use standard sizes: For example, if an 18-mil-thick MMIC carrier shim is the ideal size, but 20-mil-thick material is a standard product, try to use the 20-mil material.

- Monitor market price fluctuations: This is particularly true in the microwave industry since the price of precious metals (i.e., gold, silver, and copper) changes dramatically. For example, if the price of gold is high, avoid overplating substrate boards. Determine if a flash-plating process, which uses less gold, is thick enough for the application.

- Involve manufacturers early in design process: Some manufactures favor one technique or material over another due to equipment availability or personnel expertise. For example, it may be less expensive to replace a ribbon bond with multiple wire bonds. Knowing the strengths and weakness of manufacturers allows designers to make informed decisions.

- Push back on cost-driving requirements: Sometimes specifications are imposed that are "nice-to-haves" by the system designers but that drive up cost to the component designer. For example, if the higher-level assembly is going to be hermetically sealed, then the individual components probably do not need to be. An expensive hermetic package may be replaceable with an inexpensive plastic package.

- Push back on test requirements: The list of standard military tests covers hundreds of pages (discussed in Section 7.3). Testing and qualification is not an inexpensive process, and sometimes a duplicate test is requested at the component and higher-assembly levels. For example, it may be unnecessary to do a drop-shock test at the component level if it is planned at the system level.

• Procure long lead parts in advance: Paying above market price to expedite shipment drives up costs unnecessarily.

Practical Note

Premium capacitors that are used in radar systems are often made to order and have a 12–16-week lead time. Even the best RF capacitors only cost a few cents each. (In fact, garden-variety parts cost a fraction of a penny each.) Rather than wait until a design is finalized, it is often less expensive to purchase several values upfront than pay to expedite delivery or hold production by several months.

Exercises

1. What is the size reduction of a 15-GHz quadrature coupler on a 4-mil-thick GaAs MMIC versus a 25-mil-thick substrate with permittivity 6.0?

2. A component is only available commercially using a leaded package with 1.8-nH series inductance from the leads. What could you do to mitigate this parasitic inductance?

3. In terms of wavelength, how long is the ground return path in Figure 7.2 for a 20-GHz signal with a 4.9-mil gap? Assume that the substrate is 4-mil-thick GaAs.

4. A 200-V, 2-ns-wide voltage spike can occur on a DC bias line that feeds into an amplifier. Design a simple circuit that can be used to prevent this spike from reaching the amplifier. The nominal DC bias is 28V.

5. A 250 mil x 250 mil area of a substrate is available to add optional tuning pads to a 50-Ω microstrip line. Design a pad structure that provides maximum tunability in the space allowed. Assume that the substrate is 25-mil-thick with permittivity 6.0.

6. Under what circumstances would it be advantageous to integrate a bare MMIC as opposed to a packaged MMIC?

7. An amplifier provides 25-W output power, 30-dB gain, and 50% PAE when operated CW. Approximately what would the PAE be if operated with 10% duty cycle?

8. Sort the list of manufacturing tolerances is provided in Section 7.5.4 from most to least likely to have a significant effect. Provide an explanation as to why.

9. Binning components during production is a costly practice. From an electrical or mechanical point of view, under what circumstances might it be unavoidable?

References

[1] Defense Logistics Agency, DLA Land and Maritime Mil Specs, "MIL STD 883," Internet, http://www.landandmaritime.dla.mil/Downloads/MilSpec/Docs/MIL-STD-883/std883.pdf.

[2] Defense Logistics Agency, DLA Land and Maritime Mil Specs, "MIL-PRF-19500," Internet, http://www.landandmaritime.dla.mil/Downloads/MilSpec/Docs/MIL-PRF-19500/prf19500.pdf.

[3] Defense Logistics Agency, DLA Land and Maritime Mil Specs, "MIL-PRF-38534," Internet, http://www.landandmaritime.dla.mil/Downloads/MilSpec/Docs/MIL-PRF-38534/prf38534.pdf.

[4] Defense Logistics Agency, DLA Land and Maritime Mil Specs, "MIL-PRF-38535," Internet, http://www.landandmaritime.dla.mil/Downloads/MilSpec/Docs/MIL-PRF-38535/prf38535.pdf.

[5] Defense Logistics Agency, DLA Land and Maritime Mil Specs, "MIL-STD-1835," Internet http://www.landandmaritime.dla.mil/Downloads/MilSpec/Docs/MIL-STD-1835/std1835.pdf.

[6] Defense Logistics Agency, DLA Document Services, "MIL-STD-461," Internet, http://quicksearch.dla.mil/qsDocDetails.aspx?ident_number=35789.

[7] Richards, M., J. Scheer, and W. Holm, *Principals of Modern Radar—Basic Principles,* Edison, NJ: SciTech Publishing, 2010.

[8] Keysight Advanced Design System, http://www.keysight.com/en/pc-1297113/advanced-design-system-ads.

[9] Kalpakjian, S., and S. Schmid, *Manufacturing Processes for Engineering Materials,* Upper Saddle River, NJ: Pearson Education, Inc., 2003.

[10] United States Environmental Protection Agency, "Municipal Solid Waste Generation, Recycling, and Disposal in the United States: Facts and Figures for 2010," Document EPA-530-F-11-005, December 2011.

8

Transmit/Receive Module Integration

Chapter 7 presented techniques for ensuring that components work as expected, that they are packaged properly, that they function under necessary operating conditions, and that they are manufacturable. In a perfect world, all components could be integrated together in "plug and play" fashion without issues. Unfortunately with RF systems, it's not that simple. The effects of connecting components with mismatched impedances have already been discussed, but that's only the beginning.

When integrating components, all electrical, mechanical, and thermal interactions must be considered. Excess heat generated in one component, for example, can change the behavior of the entire module. If all components are not properly grounded or shielded, noise and oscillations are prone to dominate performance. The risk of this happening is exacerbated whenever digital circuitry is added to the mix. All of these factors must be considered along with the variation expected from manufacturing tolerances.

All RF components radiate energy in one form or another. Unless preventative measures are taken, this energy can couple onto other components and affect their performance. Loss can increase, output power and linearity can degrade, and in some cases, components can be permanently damaged. Fortunately, through the good integration practices discussed in this chapter, these issues can be prevented.

8.1 Integration Techniques

The word module is not a universally well-defined term, which can lead to confusion. One engineer's module is another engineer's subsystem, higher-level assembly, component, or RF enclosure, to name a few. Essentially, a module is any RF assembly that includes a substrate contained within a housing with at least one external connector.

Alumina is a widely used RF substrate because it offers low loss and good thermal conductivity. It is also durable and easy-to-use. Alternatively, there are many composite materials available with a wide range of electrical and mechanical properties. Most composites are either hydrocarbon, glass-filled polytetrafluoroethylene (PTFE), or ceramic-filled PTFE [1].

Housings are usually made of aluminum (for low weight and cost) or an iron-nickel-cobalt alloy (i.e., Kovar™) to minimize thermal expansion. Brass alloys (principally copper and zinc for wear and harsh environment resistance) and bronze (principally copper and tin for metal fatigue and harsh environment resistance) are also used. Brass has the advantage of being easier to machine than bronze.

RF connectors are available to accommodate a wide range of frequencies, sizes, and power-handling requirements. Chapter 9 discusses connectors. This section presents techniques for integrating components to create a module.

8.1.1 Physical Transitions

Section 2.2.1 discusses the effects of impedance mismatch. Essentially, the power that enters a component is related to the incident power and the reflection coefficient according to:

$$P_{in} = \left(1 - |\Gamma|^2\right) P_{inc}$$

(8.1)

where P_{in} is the power entering the component (W), Γ is the reflection coefficient, and P_{inc} is the incident power (W).

Improving transitions from component to component starts by minimizing this reflection coefficient. This can be done by selecting components that are well-matched at their ports (usually to 50Ω) or by designing an impedance-matching circuit (discussed in Chapter 2).

Transitioning between different types of transmission lines can also be a source of loss. The most common transition is between CPW and microstrip. CPW is a popular choice because it allows for topside RF probing and has no dispersion, as shown in Figure 8.1. Microstrip has the advantage of smaller size since the ground plane is on the backside. Other differences are discussed in Chapter 2.

To connect these structures, the ground plane can be flared as shown in Figure 8.2. This softens the discontinuity and reduces parasitic effects. The structure is symmetric so it can be used as a microstrip to CPW transition as well.

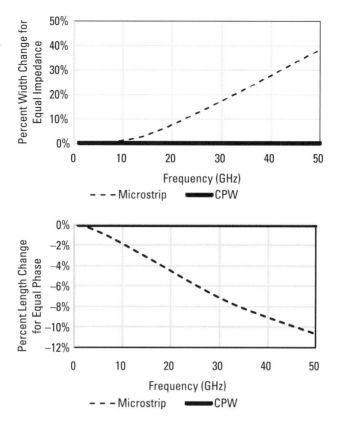

Figure 8.1 Dispersion properties of microstrip and CPW: transmission line length change to maintain phase (left) and width change to maintain impedance (right).

Figure 8.2 CPW-to-microstrip transition with reduced parasitic effects.

There are several connector options for bringing DC into a module (discussed in Chapter 9). DC feedthroughs are available that incorporate a shunt capacitor. When screwed into a metal housing, the feedthrough provides a capacitance to ground. This filters the DC input before entering the module. These are known as *feedthrough capacitors,* and they are commercially available well into millimeter-wave frequencies.

8.1.2   Wire and Ribbon Bonding

When components are integrated that do not have package leads (like those shown in Figure 7.6), wires or ribbons are used to connect the topside metal pads. Wires are usually 1 or 2 mils in diameter and ribbons are 5–10 mils in width. Chapter 3 provides (3.6), (3.7), and (3.8) for determining the equivalent inductance of a wire. Multiple wires can be used in parallel to handle the current level and reduce inductance. A single ribbon can be used to replace several parallel wires. For brevity, this text will refer to all metal bonding structures as "wires."

When connecting two transmission lines of unequal width, wires should always be placed such that they flare outward as shown in Figure 8.3. This will ease the step discontinuity to reduce loss and parasitic effects.

Bond wires must have some height in order to provide strain relief. When the temperature is cold, these small metal structures will shrink and break if adequate slack is not provided. However, if they are too long, the added inductance will affect performance. Balance must be achieved.

Alternatively, if bond wires can be placed consistently from unit to unit, then the inductance can be compensated using either a series capacitor or set of transmission lines. For example, at 5 GHz, a 1-pF series capacitor can compensate for a 1-nH wire bond. The same can be accomplished with a 50-Ω, 36-degree transmission line in series with a 68-Ω, 90-degree transmission line.

8.1.3   Proper Grounding

All too often, grounding concerns at the module level are overlooked. Components are attached to a common conductive surface (i.e., the metal housing),

Figure 8.3   Proper wire placement for transitioning between transmission lines of unequal width: orthogonal view (left) and top view (right).

and the ground plane is presumed to be equipotential everywhere. This is rarely the case. For current to flow, there must be a closed loop. A ground plane provides a low-resistance path for current to return to the source [2]. Since current-carrying conductors experience a voltage drop, it is expected that the voltage potential will vary across the surface of a ground plane. Although the potential may be small, in a noisy system it will not be negligible. For example, small fluctuations in the gate voltage of an amplifier can be detected at the output, especially for high-gain amplifiers.

The worst type of ground architecture is known as *common ground* where all components connect to the ground plane at a single point. Instead, the path to ground from a component should be as physically short (ideally less than $\lambda/20$) and as low-inductance as possible. A distributed or multipoint ground architecture is particularly important at high frequency. Figure 8.4 shows common and distributed ground designs.

In modules that integrate digital, analog, and RF, each of those circuits should have its own ground plane to prevent noise coupling. Extremely noisy circuits, like relays or motors, must share a separate ground plane. RF chokes should be added to provide isolation between circuits and ground planes.

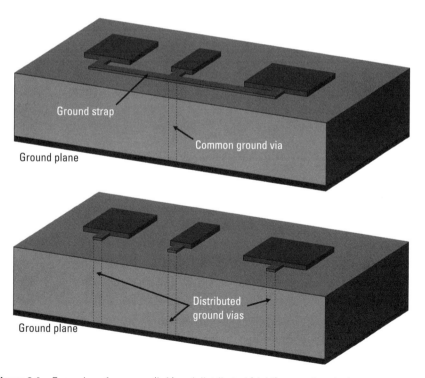

Figure 8.4 Examples of common (left) and distributed (right) grounding designs.

8.1.4 Achieving Compact Size

Size reduction is nearly always a design requirement. When integrating multiple components together into a module it is important to be mindful of two things, described as follows:

- Coupling between components: If components are placed too closely together, their fields will couple and isolation will be reduced. Figure 8.5 shows the isolation between parallel 35-degree-long microstrip lines as separation increases (in terms of substrate thickness).

  To achieve 20-dB isolation, separation must be at least four substrate thicknesses wide. A common rule of thumb is to always maintain at least three to five substrate thicknesses between components. Alternatively, components can be placed perpendicular to each other. Figure 8.6 shows the isolation between the same microstrip lines from Figure 8.5, but placed perpendicular to each other. The isolation of one substrate

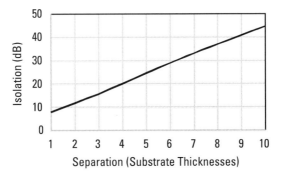

Figure 8.5 Isolation between parallel 35-degree-long microstrip lines as separation increases.

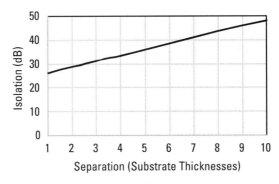

Figure 8.6 Isolation between perpendicular 35-degree-long microstrip lines as separation increases.

thickness placed perpendicularly is the same as 5.5 substrate thicknesses placed in parallel.

- Cavity resonances: All enclosed metal structures have a cavity resonance. This is the frequency where waveguide effects start to occur. These frequencies can be calculated from (2.82) in Section 2.3.6 or simulated using a 3D simulator. The lowest resonance of a module can be approximated by [2]:

$$f_{r,min} = \frac{212}{l} \qquad (8.2)$$

where f is the resonant frequency (MHz), and l is the largest dimension (m).

For example, at 10 GHz, a module should be no more than 2.12-cm-long. If it must be, then the inside of the module should be compartmentalized with conductive walls to reduce the cavity length.

8.1.5   Component Placement

Placing components near a discontinuity can increase the parasitic effects. Figure 8.7 provides an example of this. In Figure 8.7, a series element (in this case a capacitor) is placed near a mitered microstrip 90-degree corner. As that element is moved away from the miter without changing the physical length of the transmission line (left image in Figure 8.7), the insertion loss changes dramatically (right image in Figure 8.7). The miter discontinuity provides an inductance that resonates with the series element (in this case, as an LC circuit). When electrical distance is placed between those reactive components, the resonance (near 20 GHz in Figure 8.7) is alleviated.

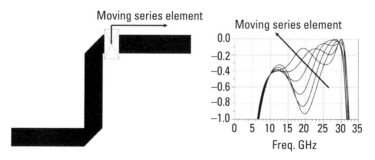

Figure 8.7   Layout of microstrip transmission line with series element moving away from the mitered corner (left) and the S_{21} trend as the element moves (right).

8.2   Preventing Oscillation

One of the most frustrating issues to overcome when designing or integrating active components is oscillation (unless, of course, you are designing an oscillator). Any active circuit has the potential to oscillate. This potential increases as gain increases. This section discusses various types, sources, and corrective actions for oscillation.

8.2.1   Even-Mode Oscillation

Even-mode oscillation can happen whenever the transistor in an amplifier is not unconditionally stable. It is caused by signals at the output coupling or feeding back to the input. It can happen without any RF applied, and the frequency is usually below the operating band. Even-mode oscillation can be determined using stability analysis (discussed in Section 4.2.1) at each stage in the amplifier chain (as opposed to analyzing the overall input and output module ports).

These oscillations are often fixed by changing DC bias capacitance or resistance values (or adding resistance to the gate bias). It is important that bias circuitry is included in stability analysis for this reason.

8.2.2   Odd-Mode Oscillation

Odd-mode oscillation can happen when multiple transistors or amplifiers are combined asymmetrically. This can occur if transistors have different electrical behavior (i.e., different I_{DSS}, pinch-off voltage, or transconductance). It can occur in amplifiers if matching conditions are different due to tolerances or unequal coupling to other components. It generally happens under RF drive, and the frequency is usually in or above the operating band. Behavior usually changes with bias or input power level.

These oscillations are often avoided by tightly matching transistor or amplifier performance. They can be fixed by adding resistors between amplifier gates or drains to improve isolation. Unfortunately, this will reduce gain, output power, and efficiency.

8.2.3   Spurious Oscillation

Spurious oscillation is a catch-all term for any frequency that appears at the output that is not a harmonic, mixing product, or cavity resonance. The origins are often unknown and cannot be predicted from stability analysis. They can happen in or out of the operating frequency range. Sometimes the oscillations disappear at certain biases or temperatures.

These are difficult to fix, especially in broadband circuits. Usually they are resolved through trial and error on the measurement bench. Moving absorbing material or metal shims (i.e., aluminum foil) around the module can sometimes reveal a way to dampen the oscillation. Adding an LC circuit to ground from the DC bias can sometimes help alleviate a spurious oscillation.

If digital circuits are within the module, noise generated from those components can contribute directly or indirectly (through harmonics or mixing) to oscillations. Section 8.7 discusses this phenomenon.

8.2.4 Ground Loops

Ground loops can occur whenever there is a disconnect in the ground plane. As mentioned in Section 8.1.3, the voltage across a ground plane may not be equipotential. By Faraday's law of induction, this can induce a current (called *ground-loop current*). This is shown in Figure 8.8.

When the shield of a coax is connected to the grounded source of an amplifier (left image of Figure 8.8), that ground node can have a different voltage potential than the voltage input. This causes a ground loop. If the shield is connected directly to the ground plane (right image of Figure 8.8), there is a better chance of achieving an equipotential ground.

Ground loops can also be avoided by [3] doing the following:

- Ensuring that all paths to ground are low-inductance (i.e., larger diameter vias have lower inductance than narrow diameter vias);

- Ensuring that all paths to ground are low-resistance (i.e, filling with high-conductivity metals);

- Using a multipoint ground design and as many ground paths as can be practically implemented.

8.3 Preventing Crosstalk and Leakage

All circuit elements radiate whenever charge is moving. Electric (or capacitive) coupling occurs whenever electric fields from two or more components inter-

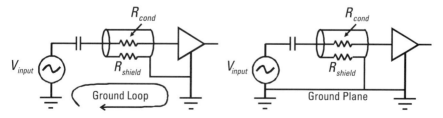

Figure 8.8 Example coax circuit with improper grounding causing a ground loop (left) and corrected circuit to prevent ground loop (right).

act. This is often mistakenly referred to as electrostatic coupling, but fields of this type are never static. Magnetic (or inductive) coupling occurs whenever magnetic fields from two or more components interact. This is commonly mislabeled as electromagnetic coupling, even though there is no influence from an electric field. Whenever both electric and magnetic fields are interacting between components, it is appropriate to classify the situation as electromagnetic.

If the distance between components is less than $\lambda/2\pi$, the fields are within the near-field region, and the effects of electrical and magnetic fields must be handled separately. If the distance between components is greater than $\lambda/2\pi$, the fields are within the far-field region, and the effects of electrical and magnetic fields can be handled as a single electromagnetic field.

The ratio of the electric field ($\vec{E}$) to the magnetic field ($\vec{B}$) is the wave impedance. In the far field, the $\vec{E}/\vec{B}$ ratio is equal to the characteristic impedance (Z_o) of the propagating medium, which can be calculated from:

$$Z_o = \sqrt{\frac{j2\pi f \mu}{\sigma + j2\pi f \varepsilon}} \tag{8.3}$$

where f is the frequency (Hz), μ is the permeability (H/m), σ is the conductivity (S/m), and ε is the permittivity (F/m). Since free space is an insulator, $\sigma \ll j2\pi f \varepsilon$ and the previous equation reduces to:

$$Z_o = \sqrt{\frac{\mu}{\varepsilon}} = \sqrt{\frac{\mu_o}{\varepsilon_o}} = \sqrt{\frac{4\pi x 10^{-7}}{1/36\pi x 10^{-9}}} = 120\pi \tag{8.4}$$

The far-field free-space wave impedance is $120\pi\,\Omega$ (approximately $377\,\Omega$). Notice that this quantity is independent of frequency.

In the near field, the $\vec{E}/\vec{B}$ ratio is dependent upon the nature of the source and the distance to the source. If the source is predominantly magnetic, the source will have high current and low voltage, and $\vec{E}/\vec{B}$ will be less than $377\,\Omega$. If the source is predominantly electric, the source will have low current and high voltage, and $\vec{E}/\vec{B}$ will be greater than $377\,\Omega$.

Additionally in the near field, the electric field attenuates at a rate of $(1/d)^3$ and the magnetic field attenuates at a rate of $(1/d)^2$, where d is the distance from the source. The wave impedance increases as d increases and approaches $377\,\Omega$ when d is $\lambda/2\pi$ (the start of the far-field region). In the far field, both electric and magnetic fields attenuate at a rate of $1/d$ [2].

Conceptually, crosstalk and leakage can be reduced by the following [2]:

• Reducing the source of the noise;

- Reducing the conducting or coupling channel;
- Desensitizing the receptor.

Generally this can be accomplished by either shielding, properly grounding, filtering, separating components, or reorienting components. All of these techniques are discussed in this chapter.

8.3.1 Electric Coupling

If we assume that a component can be approximated as a perfect conductor with regard to radiation, then electric coupling analysis is much more straightforward. Between any two components, the voltage that couples from component 1 to component 2 can be calculated from [2]:

$$V_C = V_1 \frac{j2\pi f\left(\dfrac{C_{12}}{C_{12}+C_{2G}}\right)}{j2\pi f + \dfrac{1}{R_{2G}\left(C_{12}+C_{2G}\right)}} \tag{8.5}$$

where V_C is the voltage coupled to component 2 (V), V_1 is the voltage on component 1 (V), f is the frequency of V_1 (Hz), C_{12} is the capacitance between component 1 and component 2 (F), C_{2G} is the capacitance between component 2 and ground (F), and R_{2G} is the resistance between component 2 and ground (Ω). V_1 can be the desired signal or an undesired noise voltage.

The capacitance between two parallel conductors (which we are approximating our components to be) can be calculated by [2]:

$$C_{12} = \frac{\pi \varepsilon}{\cosh^{-1}\left(\dfrac{D}{d}\right)} \tag{8.6}$$

where ε is the permittivity of the volume between components ($\varepsilon = \varepsilon_o \varepsilon_r$), D is the distance between components (m), and d is the diameter of the component (m).

In most cases, R_{2G} is much less than $[j2\pi f(C_{12} + C_{2G})]^{-1}$, so the V_C equation can be simplified to:

$$V_C = V_1 j2\pi f R_{2G} C_{12} \tag{8.7}$$

Equation (8.7) indicates that the coupled voltage is proportional to the source voltage, frequency, the resistance of component 2 to ground, and the capacitance between components. Since the source voltage and frequency usually cannot be changed, the only ways to reduce coupling are to improve (reduce the resistance) the connection from component 2 to ground or to decrease the capacitance between components. The latter can be accomplished by doing the following:

- Reorienting the components (i.e., perpendicular or off-angle instead of parallel);
- Adding shielding (discussed in Section 8.3.3);
- Increasing distance between components (8.6).

In the case where R_{2G} is much larger than $[j2\pi f(C_{12} + C_{2G})]^{-1}$, the V_C equation can be simplified to [2]:

$$V_C = V_1\left(\frac{C_{12}}{C_{12} + C_{2G}}\right) \tag{8.8}$$

Three observations about the coupling voltage can be made from (8.8):

- It is independent of frequency;
- The magnitude is larger than when R_{2G} is small;
- It is a function of the capacitance between conductor 2 and ground.

8.3.2 Magnetic Coupling

We know from the Ampère-Maxwell law (1.16) that an electric current produces a circulating magnetic field. This magnetic field can couple from one object to another. If a current flow in component 1 induces a magnetic flux in component 2, a mutual inductance between the components can be calculated from [2]:

$$M_{12} = \frac{\phi_{12}}{I_1} \tag{8.9}$$

where M_{12} is the mutual inductance, ϕ_{12} is the flux in component 2 due to component 1, and I_1 is the current in component 1. M_{12} is the magnetic analog to the electric capacitance C_{12}.

The mutual inductance between two parallel conductors (which we are approximating our components to be) can be calculated by [2]:

$$M_{12}\left(\frac{\mu H}{in}\right) = 0.0025 \ln\left[1 + \frac{(2h)^2}{D^2}\right] \tag{8.10}$$

where h is the height above the ground plane (in) and D is the distance between components (in).

The coupling voltage induced from a magnetic field can be derived from Faraday's Law (1.13):

$$V_C = -\frac{d}{dt}\int_S \vec{B} \circ \hat{n}da = j2\pi f M_{12}I_1 \tag{8.11}$$

If the flux density is constant over the area of the surface and varies sinusoidally with time, (8.11) reduces to [2]:

$$V_C = j2\pi f BA\cos\theta \tag{8.12}$$

where B is the rms value of the flux density (Teslas), A is the area of the magnetic loop (m^2), and θ is the angle between the components (radians).

The magnetic flux density for a point in space a distance from a current carrying conductor can be calculated from the Biot-Savart law [2]:

$$B = \frac{\mu I}{2\pi r} \tag{8.13}$$

where μ is the permeability of the volume between components ($\mu = \mu_o = 4\pi \times 10^{-7}$ H/m for free space), I is the current in component 1, and r is the distance between components (meters).

Equations (8.11) and (8.12) indicate that the coupled voltage is proportional to the source current, frequency, flux density, and inductance between components. Since the source current and frequency usually cannot be changed, the only ways to reduce coupling are to reduce the flux density or decrease the inductance between components. This can be accomplished by doing the following:

- Reorienting the components (i.e., perpendicular or off-angle instead of parallel) to change θ;

• If the source of the current is a wire, coiling (or twisting) the wire to cancel the magnetic field and reduce B;

• Increasing distance between components to decrease M_{12}.

Air core or open magnetic core inductors are prone to magnetic coupling since their flux extends well outside the volume of the component. Particular attention should be paid to either shield these components or space them away from other components. Alternatively, closed magnetic core inductors can be used.

8.3.3   Shielding

Shielding introduces an electrically conductive partition that surrounds a volume to provide isolation to the outside. When placed around a noise or coupling source, they are effective at containing (or blocking) electromagnetic fields. When the shield is grounded, the coupled voltage is shorted to ground and the component within the shield is isolated. A proper shield should be grounded at a distance no less than every $\lambda/20$.

In high-gain power amplifiers, it is critically important to ground the shield properly. Otherwise, coupling between the final-stage output and the shield can cause a feedback loop to couple between the first-stage input and the shield. This can be prevented by grounding the shield to the amplifier common terminal (usually the source since most RF amplifiers use a common-source configuration). In most cases, the amplifier common terminal is directly connected to the ground plane so that the shield can also be connected directly to the ground plane. In cases where the amplifier has an isolated ground (i.e., a floating ground to prevent noise coupling to a sensitive receiver circuit), the shield should connect to the amplifier ground and remain isolated from the system ground (or Earth ground).

Shielding can be effective for magnetic coupling, but it's not as straightforward as preventing electric coupling. Magnetic shielding functions by inducing a ground-return current in the direction opposite to the main propagating signal. The two currents generate magnetic fields in directions opposite to each other, which effectively cancels them out. The difficulty in implementing this approach is getting the ground locations on the shield correct. If the opposing magnetic field is not equal, opposite, and in alignment with the main magnetic field, the shield will be ineffective (or will only be effective over a small range of frequencies). Using a coaxial cable (discussed in Chapter 9) can provide a well-controlled environment for implemented a high-frequency magnetic shield as long as both ends of the coax are grounded to implement the ground-return loop. At low frequency, coaxial cables are ineffective at magnetic shielding. A twisted-pair cable is much more effective, and the number of twists per unit length should be maximized for peak effectiveness.

Instead of shielding the receiving component, placing a well-grounded shield over the radiating component can also be effective. Regardless of the shield location, it can only truly be effective if there are no gaps, holes, or breaks in the shield.

The effectiveness of a shield can be determined by adding the absorbed loss in the shield with the loss from reflection. At low frequency for electric fields, reflection loss dominates, and at high frequency absorption loss dominates. Absorption loss dominates magnetic fields at all frequencies.

The absorption loss of a shield depends on the thickness and skin depth and can be calculated by [2]:

$$A = 20 \frac{t}{\delta_s} \log e \qquad (8.14)$$

where A is the absorption loss (dB), t is the thickness of the shield (m) and δ_s is the skin depth (m) (2.87). Absorption loss is the same whether in the near field, far field, electric field, or magnetic field.

The reflection loss can be calculated from [2] assuming that:

$$R = 20 \log \left(\frac{120\pi}{4\sqrt{\frac{2\pi f \mu}{\sigma}}} \right) \qquad (8.15)$$

where R is the reflection loss (dB), f is the frequency (Hz), μ is the shield material permeability (H/m), and σ is the shield material conductivity (S/m). From the perspective of preventing coupling to the outside, higher reflection loss is better than lower reflection loss—although it is true that too much reflection can couple back to the circuit and change performance.

The alternative to using a shield material with good conductivity is to use one with high permeability (that is, a magnetic or ferrite-loaded material). Since skin depth is inversely proportional to permeability, increasing μ also increases absorption loss. This makes the shield more effective, particularly at low frequency. Concurrently, reflection loss is inversely proportional to permeability, so increasing μ reflects less energy, which may be beneficial for some sensitive components that cannot tolerate reflections.

Unfortunately, magnetic shielding is not as effective for low-frequency electric fields. As a compromise, the inside of a conductive shield can be lined with high-permeability material. This stack-up is more costly than using one of the other methods, but it does offer broader-band shielding.

The best shield completely covers the component being shielded, but unfortunately that is not always possible. Openings may need to be added to feed in bias lines, RF lines, or cooling structures. These openings should be used minimally. Some guidelines for minimizing the effect of shield openings include the following [2]:

- Multiple small openings are better than one large hole of the same area;
- Keep the openings to less than $\lambda/20$ each;
- Space openings by more than $\lambda/2$;
- Distribute openings to different sides of the module.

8.4 Thermal Considerations

When integrating multiple components, it is important to determine the change in temperature due to the surrounding components. Section 4.5.1 discussed the equations for performing thermal analysis, although thermal analysis is usually determined through simulation. Components that generate large amounts of heat (i.e., high-power amplifiers) may need to be placed away from temperature-sensitive components (i.e., LNAs).

Generally, it is desirable to pull heat away from the sources (i.e., transistors) as quickly as possible to prevent overheating, loss of performance, and reduced lifetime. Reducing the thermal resistance to the heat sink can be accomplished using the following:

- High thermal conductivity ceramics and semiconductors: The appendix lists the thermal conductivities of many common materials, including aluminum nitride (285 W/m·K) and silicon carbide (114 W/m·K);
- High thermal conductivity metal carriers: The appendix lists the thermal conductivities of many common materials, including aluminum (235 W/m·K) and copper (401 W/m·K);
- CVD diamond: A relatively new addition to the microwave market, CVD diamond has thermal conductivity up to 1,800 W/m·K. Due to the high price, it is usually placed only under the hottest components to effectively remove heat from the source;
- Thermally conductive pastes, gels, greases, and films: These materials are thinly spread between the main conducting materials (i.e., substrate and housing) to fill in any air gaps (air is nearly 9,000 times less conductive than aluminum).

It is also desirable to spread the heat around the module surface and eliminate it from the module as much as possible. There are several ways to do this, described as follows.

- Heat pipes: As the name suggests, these are tubes filled with a material that is superconducting in one direction (in this case, down the pipe). They can be placed parallel or perpendicular to the heat source(s) and act like a thermally conductive superhighway to move heat from the source to the sink. These are particularly useful in applications where the heat sink is located far from the heat source.

- Phase change materials: Substances added to the module that are designed to melt at the upper temperature limit. When that temperature is exceeded, the substance changes phases (i.e., melts or evaporates) and absorbs heat in the process. This process cools the module. Materials are available that can be effective to more than 150°C.

- Convection cooling: Forcing fluid (i.e., air, deionized water, or glycol) over a module can increase the rate of heat removal by 10 times (discussed in Section 4.5.1).

8.5 Mechanical Considerations

When components (especially hybrid circuits) are integrated, it is important to pay attention to the type of metal used in the carrier. Whenever two different metals touch, there is a risk of galvanic action taking place. This is especially true in the presence of moisture or water vapor. Table 8.1 lists common metals used in RF modules from most anodic to most cathodic.

The further away from each other the metals reside in Table 8.1, the greater the galvanic reaction between them (i.e., the greater the voltage generated). Whenever possible, only materials within the same group should be used together. Otherwise, one or both metals will corrode and eventually the module will cease to function.

Practical Note

Attaching copper (group 4) to aluminum (group 2) is popular because it combines the thermal spreading of copper with the low cost and light weight of aluminum. Together, they generate a serious galvanic reaction. and eventually the aluminum will corrode away. This reaction can be slowed if the copper is coated with lead-tin solder (group 3) since the galvanic reaction between aluminum and lead-tin is much less [2].

Table 8.1
Common Module Metals Listed from
Most Anodic to Most Cathodic [2]

Group	Material
1	Magnesium
2	Zinc
2	Aluminum
2	Cadmium
3	Steel
3	Iron
3	Lead-tin solder
3	Lead
3	Tin
4	Nickel
4	Brass
4	Copper
4	Bronze
4	Copper-nickel
4	Silver solder
4	Silver
5	Graphite
5	Gold
5	Platinum

8.6 Module Simulation and Monte Carlo Analysis

Simulating at the module level can be cumbersome. Each component is probably simulated by a different designer, and integrating multiple design files together can be nontrivial. Additionally, components that are purchased commercially may not have behavioral models. Despite these challenges, the likelihood of achieving first-pass success is significantly higher when higher-level assembly simulations are performed.

Keysight ADS is equally effective at simulating modules as it is at simulating components. Each design file can be represented as a single component with multiple ports for DC and RF input and output. Alternatively, a measurement file can be used when a model is not available. Multiple design or measurement components can be combined to create higher-level components, and this process can be repeated until the entire module can be represented in a manageable form. Once this main simulation file is created, module-level performance can be simulated and compared against the specification. Sources of losses can be determined. Stability and crosstalk analysis can also be performed.

The one limitation of this "master" simulation file is that it assumes that all mechanical dimensions, assembly alignment, and device performance are perfect. In reality, we know this is not possible. Machine and circuit board shops and assembly houses typically hold tolerances to within a few mils. Devices generally fall within ±10% of the nominal value. A small unexpected and undesired parasitic can have a significant impact on performance. Tables 8.2 and 8.3 list the reactance (imaginary part of impedance) created by an undesired series inductance and capacitance at 5, 20, and 50 GHz.

Tables 8.2 and 8.3 show that it doesn't take much to create a large reactance, and this causes two concerns:

- How does this variation affect performance?
- How does this variation affect stability?

Arguably, the second concern is the most important. A design can have ample margin over the specification, but that doesn't matter if the component fails under RF drive due to instability.

To determine how much variation can be expected due to tolerances, a Monte Carlo analysis can be performed. Simulators allow a designer to set a

Table 8.2
Reactance For Small-Series Inductance at Three Frequencies

Undesired Series Inductance	5 GHz	20 GHz	50 GHz
0.1 nH	3.14Ω	12.57Ω	61.42Ω
0.25 nH	7.85Ω	31.42Ω	78.54Ω
1 nH	31.42Ω	125.66Ω	314.16Ω
5 nH	157.08Ω	628.32Ω	1570.80Ω

Table 8.3
Reactance for Small-Series Capacitance at Three Frequencies

Undesired Series Capacitance	5 GHz	20 GHz	50 GHz
0.1 pF	−318.31Ω	−79.58Ω	−31.83Ω
0.25 pF	−127.32Ω	−31.83Ω	−12.73Ω
1 pF	−31.83Ω	−7.96Ω	−3.18Ω
5 pF	−6.37Ω	−1.59Ω	−0.64Ω

nominal value, range of possible values, and type of variation (usually uniform or Gaussian) for each physical variable (i.e., length, width, thickness, and placement). The simulator will vary the parameters pseudorandomly equally across the range specified (if the variation type is set to uniform) or weighted more heavily around the nominal (if the variation type is set to Gaussian). Each set of parameters simulated is called a *trial*. A good rule of thumb for thorough coverage is to determine the minimum number of trials per:

$$N_{trials} = \left(N_{param}\right)^e \quad \text{for } N_{param} \geq 2 \tag{8.16}$$

where N_{trials} is the number of Monte Carlo trials, N_{param} is the number of parameters varied, and e is the natural log (≈ 2.718). Figure 8.9 plots this equation.

As the number of parameters gets large, the suggested number of trials can reach tens or hundreds of thousands. It may be necessary to perform a full Monte Carlo analysis on each component, determine the range of possible performance, create a new behavioral model that is hard-coded to this range of possible performance (ADS has built-in behavioral models for components that makes this straightforward), and use that simplified component in the module-level Monte Carlo analysis. For example, a power amplifier may have 35 variables, so it needs to run approximately 16,000 trials. From that analysis, a range of output power, efficiency, linearity, noise, and S-parameters will be determined. Those ranges can be entered into the ADS "Amplifier2" component. Now there are only eight or so variables that are needed to provide the same behavior variation in the next higher-level analysis.

Figure 8.10 shows the results of an example Monte Carlo analysis for a two-way Wilkinson divider. The case where all variables are as designed is

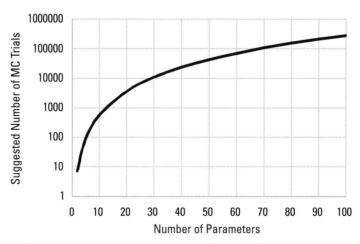

Figure 8.9 Suggested number of Monte Carlo trials (log scale) for a given number of parameters (8.16).

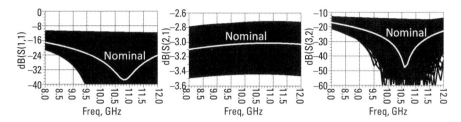

Figure 8.10 Example Monte Carlo analysis for a two-way Wilkinson divider.

labeled the nominal. Notice the spread of performance that can be expected due to electrical, mechanical, and assembly tolerances.

8.7 Incorporating Digital into an RF Module

As digital circuits become smaller, more powerful, easier to program, and less power consuming, incorporating them into RF modules has become more popular and for good reason. ADCs sample an RF waveform and convert it to a digital signal. Once in the digital domain, software can be written to perform math or logic functions on the data. Complicated functions are usually much easier to implement in the digital rather than analog domain. DACs generate an RF waveform from the digital signal.

A digital integrated circuit that can be programmed (and reprogrammed) by the designer is called a *field programmable gate array* (FPGA). FPGAs have a library of math functions (i.e., SIN, EXP, SQRT) and logic functions (i.e., AND, OR, XOR) that can be used by the programmer to analyze or manipulate data. In comparison, an *application specific integrated circuit* (ASIC) is hardcoded to perform a specific function and cannot be changed by the user. Since they are fully optimized for one purpose, ASICs can outperform FPGAs programmed to fulfill the same purpose. However, advancements in FPGA technology are closing this performance gap, and ASICs are becoming less popular.

Unfortunately, adding a digital circuit to an RF module is not without challenges. High-speed circuits have switching or clock speeds in the hundreds or thousands of megahertz range. These frequencies may be in the RF band (direct coupling) or have harmonics or mixing products that are in the RF band (indirect coupling). Signals can radiate from one component to another or propagate through the ground plane to the RF circuitry (if digital and RF circuits share a ground plane).

Issues related to digital circuits can be difficult to trace back to their origin. This section discusses techniques for using digital circuits effectively in RF modules.

8.7.1 Common Digital Uses

Digital circuits are most commonly used to add "smarts" to RF circuits. They can collect data from sensors (i.e., temperature probes or chemical detectors), bias circuits, and input/output waveforms and perform analysis. Simple circuits may adjust bias voltages to compensate for temperature effects. More complicated circuits may correct for aging, radiation, or other environmental effects.

Circuits are now available that can process at RF (gigabit) speed, and microwave speed is on the horizon. Processing at this speed enables real-time waveform sampling, manipulation, and adaptability. Bias and port impedance can be optimized to provide peak performance (i.e., output power, efficiency, and linearity) to the system in real time [4]. If advanced knowledge of the performance need is available (as is the case in a *cognitive radar*), the digital circuit can adapt the RF circuitry so the module performance is modified exactly when needed [5].

APD, discussed in Section 4.4.3, is a way to improve amplifier linearity. The same technique can be implemented using digital means. This is referred to as DPD. Since DPD is implemented using a single integrated circuit, it can be much more versatile than APD. For example, an FPGA can be programmed to implement DPD and correct for performance changes due to bias drift, temperature, aging, and waveform characteristics (i.e., channel width). The same can be implemented using APD, but it would require an extensive and complicated set of hardware. Additionally, DPD can be programmed to calibrate each module at start-up. This would correct for module-to-module variation due to manufacturing tolerances. An APD circuit would either need to be calibrated during manufacturing (a costly practice) or designed to fit nominal performance (some units will perform better than others). When only a relatively small improvement to linearity is needed (10–15 dB), adding a small APD circuit would be easier than incorporating a digital circuit.

Digital circuits are great for implementing built-in test (BIT), built-in self-test (BIST), or built-in test and repair (BITR) capability. Usually couplers are placed at strategic points in the module (i.e., at the input and output of an amplifier). The coupled port is connected to an FPGA through an ADC. A simple test circuit determines whether the signal amplitude is as expected. If not, then a failure has been detected. An error message can be transmitted to the user (if the infrastructure is in place to do so), or an auxiliary set of hardware can be enabled (if available). The latter capability (repair) is an example of BITR. If performance is degraded, a BITR circuit may be able to modify circuit parameters (i.e., bias voltage or tunable elements) to regain performance.

Digital circuitry can also implement modulation schemes. Pulse-width modulation (PWM), for example, is used to reduce average prime power consumption. PWM is mostly used in CW applications, but the same principal applies to pulse modulators commonly used in past and present radars. Figure 8.11 demonstrates the operation of PWM.

In Figure 8.11, the analog waveform is a simple sine wave (dotted curve). The amplifier is turned on in short pulses (or *bursts*, light color trace) and driven into an averaging filter (i.e., an inductive circuit). The modulated output (solid curve) then resembles a stepped version of the analog waveform. Since the amplifier is only consuming prime power while pulsing, most of the time no prime power is being consumed. In order to maintain an acceptable level of signal fidelity, high-speed digital circuitry is needed.

8.7.2 Current Digital Infrastructure

In most (if not all) modern radars, digital circuitry already exists. Figure 8.12 shows a high-level block diagram for a modern radar.

All of the "brains" in a radar lie in the digital circuitry. Digital circuits determine the appropriate radar waveform characteristics in the transmitter. In the receiver, they process the data into information that is valuable to the operator. Since antennas cannot propagate digital signals, the purpose of the RF circuitry is to prepare the waveform for propagation (i.e., through frequency upconversion and amplification), propagate and receive the waveform (i.e., through the antenna), and prepare the waveform for analysis (i.e., through amplification, frequency downconversion, and demodulation). If digital circuitry provides the brains, RF circuitry provides the "brawn." Since digital circuitry is already available in the system, RF designers might as well use it.

System-on-chip (SOC) is an approach that integrates digital, analog, and RF circuitry onto a single chip. It is popular in the nonradar commercial market as a means to keep up with the demands for size reduction (i.e., smaller cell phones and thinner televisions). An entire radar T/R module (minus the antenna) can be reduced to a single integrated circuit. An impressive degree of size and cost reduction can be achieved using SOC. There are significant design challenges, including preventing crosstalk, maintaining ground plane integrity, and heat removal.

8.7.3 Digital Radiation

High-speed digital circuits can have clock speeds in the hundreds to thousands of megahertz range. These signals can radiate or couple to other components that can radiate. The faster the switching speed, the greater the potential for radiation. Similarly to RF, radiation from digital circuits exhibit near- and far-field behavior.

There are two types of digital radiation. The first, differential mode radiation, is caused from conductive paths around a circuit. This effect is similar to

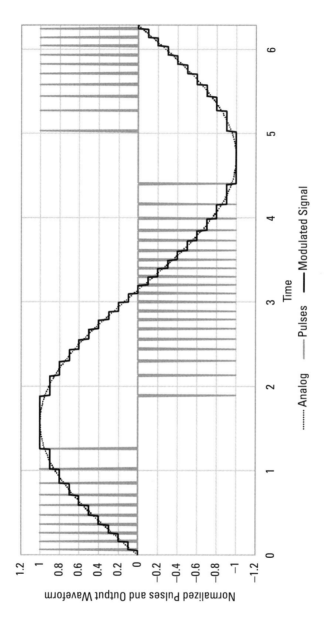

Figure 8.11 The operating principal of PWM.

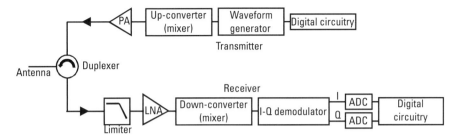

Figure 8.12   High-level block diagram of modern radar with digital circuitry.

a small loop antenna, so the magnitude of the electric field can be determined by [2]:

$$E_{dm} = 131.6 \times 10^{-16} \left(\frac{f^2 AI}{r} \right) \sin \theta \qquad (8.17)$$

where E_{dm} is the differential mode electric field (V/m), f is the frequency (Hz), A is the area of the conductive path (m²), I is the current flowing through the path (A), r is the distance measured from the component center (m), and θ is the angle measured from the component surface (radians).

Common mode radiation is caused from unequal voltage drops to ground at various locations around the circuit that spur radiation. This effect is similar to a small monopole antenna, so the magnitude of the electric field can be determined by [2]:

$$E_{cm} = 4\pi \times 10^{-7} \left(\frac{flI}{r} \right) \sin \theta \qquad (8.18)$$

where E_{cm} is the common mode electric field (V/m), f is the frequency (Hz), l is the path length to ground (m), I is the current flowing through the path (A), r is the distance measured from the component center (m), and θ is the angle measured from the component surface (radians). Mitigation techniques are discussed in Section 8.7.4.

8.7.4   Avoiding Mixed-Signal Issues

Whenever analog (or RF) and digital are used together, the combination is known as *mixed signal*. The following techniques are suggested to ensure proper mixed-signal operation:

- Don't use integrated circuits with switching or processing speed greater than what is needed. For example, temperature changes happen over many seconds (or longer); there is no need for microsecond resolution in a temperature-compensation circuit. Faster speeds increase the likelihood of a spurious signal generating in-band, consume more prime power, and are more expensive.

- Be mindful of switching speed and internal clock timing. Calculate all potential harmonics and mixing products and, if possible, add filters to remove them prior to the RF circuitry. Keep a list during testing so that if a spurious signal is detected, it may be traced to the digital circuit.

- Coupling is stronger between components whose direction of propagation is parallel. Rotating components so they are placed perpendicular to each other will reduce coupling.

- Separate and isolate RF and digital ground planes. This requires multiple power supplies but can improve isolation by 60 dB.

- Route metal traces around components rather than under. As shown Figure 8.13, when components are in the way, signal traces can route under or around the obstruction. Routing through the substrate aids in size reduction, but it also causes more radiation.

Incorporate a *frequency-selective surface* (FSS) to block unwanted frequencies. Isolation can be improved by as much as 100 dB for narrow-band applications.

An *electronic band gap* (EBG) is a type of FSS that can be patterned on the topside or backside (ground plane) of a substrate to block frequencies from propagating. An example of both topologies is shown in Figure 8.14, where select frequencies are blocked from traveling between port 1 and port 2. Example EBG structures are shown in Figure 8.15 [6].

By wrapping an EBG around a component, as shown in Figure 8.16, it is isolated from other components within the frequency range of the EBG. It is useful for blocking difficult-to-remove frequencies like those generated from digital switching speed.

Exercises

1. The front-end transmitter of a two-channel (or dual-band) radar operates across S- and X-band. Unfortunately, it has a spurious oscillation at 5 GHz, and you are tasked with resolving the oscillation. You have been granted a rare opportunity to spend a very brief moment

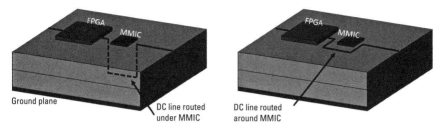

Figure 8.13 Options for routing signal lines under (left) or around (right) a component.

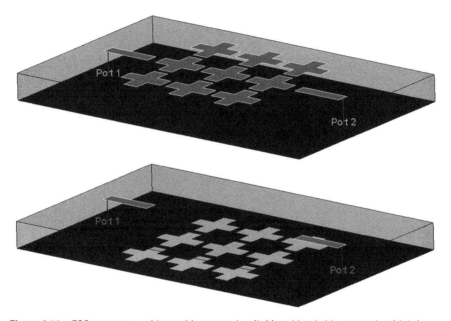

Figure 8.14 FSS structures with topside patterning (left) and backside patterning (right).

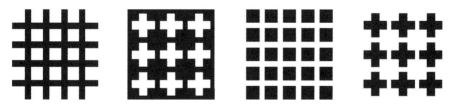

Figure 8.15 Inductive-style (left images) series LC and capacitive-style (right images) parallel LC-to-ground EBG structures.

with the chief engineer. What one question would you ask in order to confirm or rule out each of the potential sources listed as follows (one question per potential source)?

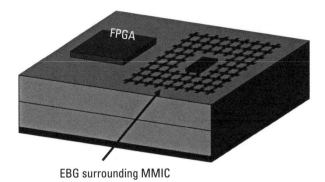

EBG surrounding MMIC

Figure 8.16   Example FPGA isolated from RF MMIC by surrounding MMIC with capacitive-style EBG structure.

- Improper grounding;
- Poor isolation between components;
- Cavity resonance;
- Even-mode oscillation;
- Odd-mode oscillation;
- Ground loop.

2. Consider the same situation as in the first exercise, except with a spurious oscillation of 9 GHz.

3. Consider the same situation as in the first exercise, except with a spurious oscillation of 50 MHz.

4. Two 15-mil-wide metallic components are spaced 15 mil apart and 10 mil from the ground plane. The entire volume is encased in liquid crystal polymer.

  - What is the capacitance between them?
  - What is the mutual inductance between them?
  - What is the absorption loss at 6 GHz of a 125-mil-thick aluminum shield?

5. Whenever aluminum carriers are plated with gold, a thin layer of nickel is usually plated in between. What purpose does this layer serve?

6. A design has 30 parameters that need to be varied as part of a Monte Carlo analysis. Rather than run ~10,500 trials (30^e trials), the designer decides to split the analysis into two steps. The first 15 parameters are run with 1,570 trials (15^e trials) and project 5% fallout. The second 15 parameters are also run with 1,570 trials and project 10% fallout.

- Can anything be concluded from this information about the expected fallout if the designer had run all 30 parameters simultaneously?

7. Is splitting a set of Monte Carlo parameters into multiple smaller sets a suitable substitution? Why or why not?

8. In Figure 8.10, it is shown that some Monte Carlo trials show performance that is better than nominal, which is supposed to be the optimized configuration. How can this be the case?

References

[1] Rogers Corporation, "Advanced Materials for PCBs, Power Distribution, Impact Protection," Internet, http://www.rogerscorp.com, 2015.

[2] Ott, H., *Noise Reduction Techniques in Electronic Systems,* New York, NY: John Wiley & Sons, Inc., 1988.

[3] Senturia, S., *Microsystem Design,* Norwell, MA: Kluwer Academic Publishers, 1940.

[4] Kingsley, N., and J. R. Guerci, "Adaptive Amplifier Module Technique to Support Multifunction Radar Architectures," *MSS Tri-Service Radar Conference,* July 2014.

[5] Guerci, J. R., *Cognitive Radar: The Knowledge-aided Fully Adaptive Approach,* Norwood, MA: Artech House, 2010.

[6] Hooberman, B., "Everything You Ever Wanted to Know About Frequency-Selective Surface Filters but Were Afraid to Ask," Internet, http://cosmology.phys.columbia.edu/group_web/about_us/memos/hooberman_filters_memo.pdf, May 2005.

Selected Bibligraphy

Kenington, P., *High-Linearity RF Amplifier Design,* Norwood, MA: Artech House, 2000.

Maas, S., *Practical Microwave Circuits,* Norwood, MA: Artech House, 2014.

Holzman, E., *Essentials of RF and Microwave Grounding,* Norwood, MA: Artech House, 2006.

9

On the Measurement Bench

For an engineer, there is probably no greater thrill than seeing a design come to life. Unfortunately, despite all the preparation required to make a low-risk design, simulations can't predict everything.

Once the design is complete, the real fun begins. The bench demonstration is the engineer's first opportunity to determine how accurate the design really was. If accurate models have been used (Chapter 3) and a proper Monte Carlo has been performed (Section 8.6), then measured results should come close to predicted performance.

This chapter is dedicated to getting the best measured results and addresses what to do when the results are not as expected. Further, the chapter discusses the importance of determining the uncertainty associated with a particular measurement or measurement bench and presents methods for designing a good test fixture and de-embedding its contribution to measured results. The chapter also lists different types of connectors, adapters, and cables with their respective advantages. Finally, the chapter describes how to stabilize active circuits (that have achieved their desired results) to maintain performance over their lifetime.

9.1   Measurement Uncertainty

Before beginning a measurement, it is important to determine the uncertainty of the test bench. In metrology (the science of measurement), there are two types of measurement uncertainty:

- Accuracy: How closely the measured result is to the true value;

• Precision: How repeatable the result is from measurement to measurement.

For example, if the output power of an amplifier is exactly 10W, but the instrumentation measures 5W after testing multiple times, the measurement is precise but not accurate.

Ensuring that a measurement setup is both accurate and precise is critically important. Otherwise, there is no way of knowing if a design meets specification. Every measurement has some degree of measurement uncertainty, which is the difference between the actual value and the one reported from the measurement. For example, if a current meter has a measurement uncertainty of ±0.1% and the meter display reads 12.5000A, the actual current is between 12.4875 and 12.5125A.

Practical Note

Metrologists avoid using the term "measurement error" when referring to measurement uncertainty. "Error" gives the impression that a mistake or oversight has been made, which isn't necessarily the case. "Uncertainty" implies a degree of randomness, which is a better description of the phenomenon.

When a component is being measured (especially for the first time) and the results are not as expected, it's natural to assume that the problem is in the component. It's always a good idea to first check the measurement bench to determine if the results are within the degree of measurement uncertainty. (Ideally, this should be done before starting the measurement so a proper expectation is set.) At a high level, measurement uncertainty can be determined by the following:

• Determining the uncertainty for the measurement equipment (usually listed in the product specification) and performing a statistical analysis to roll up the individual uncertainties into a bench uncertainty;

• Testing a measurement standard (accurately calibrated device with known performance) or golden unit (device with known good performance that other units can be compared to empirically).

Neither approach is without limitations. Equipment specifications tend to state conservative measurement uncertainty values. To achieve high test yields, engineers will be required to overdesign components to compensate for the wide uncertainty range. Measuring "known" standards can determine the *precision* of a measurement bench, but they introduce their own uncertainty when determining *accuracy*.

> **Practical Note**
>
> It is not acceptable practice to improve accuracy by measuring a component multiple times and averaging the results. Uncertainty should be presumed to be uniformly distributed so the average value is not necessarily the most accurate.

If measurement uncertainty is greater than desired, it can be minimized by the following [1]:

- Using only equipment with valid calibration (some equipment must be calibrated annually);
- Reducing cable lengths;
- Reducing the number of adapters;
- Avoiding chained adapters to achieve the necessary input and output type (i.e., if an SMA male to 2.9-mm female is needed, don't insert an additional 3.5-mm adapter because that's what is available. Procure an SMA male to 2.9-mm female adapter);
- Using the same connector type (i.e., 2.9-mm) throughout the bench;
- Using a torque wrench to connect cables, connectors, and adapters to ensure that proper tightness is provided;
- Using only quality cables, connectors, and adapters (dispose of any broken or questionable units to prevent inadvertent use by someone else);
- Add an attenuator or isolator to protect the measurement equipment from high-power reflections.

Measurement uncertainty cannot be avoided so it should always be incorporated into a specification as design margin.

9.2 Test Fixture Design

While in production, special fixtures may be used to perform component-level testing without causing wear to the interfaces (i.e., connectors, microstrip launch, or CPW launch). Production fixtures are designed for fast, repeatable (precise), and reliable (accurate) testing. Design techniques for those fixtures are outside the scope of this book, but engineering-level (or R&D-level) test fixtures are worthy of discussion [2].

A well-designed test fixture would introduce minimal loss, impedance mismatch, and parasitic effects to the component being tested [commonly referred to as the *device under test* (DUT)]. A poorly designed test fixture can actually change the DUT performance and render the measured results useless. Calibration is used to mathematically remove the effects of the measurement bench and fixtures so that the DUT performance can be isolated. However, calibration does not change what the DUT "sees" when connected. For example, a fixture with poor impedance match can be calibrated to represent a perfect 50-Ω match, but the DUT will still behave as if it is poorly matched at its ports. The measured results will be misleading.

Similarly, a good test fixture should provide the same environmental conditions as the operational environment. For example, if a DUT operates hotter in the test fixture than it would in the operating environment because the test fixture has a lower thermal conductivity, the measured results will be misleading. This can be overcome by adding a thermocouple at the DUT-to-test fixture interface and monitoring temperature.

For DUTs that do not use rugged connectors at the ports (discussed in Section 9.2.2), light contact is usually desired at all points between the DUT and test fixture. A variety of plunger-style or pogo pins are commercially available to make light, short-duration contact. Sometimes wire or ribbon bonds can provide a good connection with minimal disturbance to the DUT connection pads. Figure 9.1 shows a simple two-port DUT with ribbon-bond connections.

Practical Note

For DUTs with active components, include voltage "test points" throughout the circuit. During debug, it is very useful to have connection points where DC probes can be attached without risking damage to the RF circuitry. If these test points can be connected directly to ports on the test fixture, then gate and drain voltages (for example) can be monitored in real time during testing.

9.2.1 De-Embedding Fixture Effects

If a measurement fixture could ever be perfectly matched with zero insertion loss and phase length, then it would have no contribution to the measured

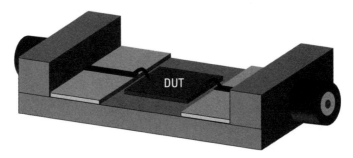

Figure 9.1 Simple two-port DUT in a test fixture with ribbon-bond connections.

DUT performance. Unfortunately, even the best measurement fixtures have some amount of impedance mismatch, loss, and phase length. In some applications (particularly below 10–100 MHz), these quantities may be negligible and can be ignored. For most applications, fixture effects must be removed (or de-embedded) from the measured results.

Arguably the easiest and most accurate way to de-embed fixture effects is to create an equivalent circuit model between the DUT and each measurement port. Figure 9.2 shows a circuit for the fixture shown in Figure 9.1. Since the measurement fixture is symmetric, there is a line of symmetry in the circuit as well.

A frequency-dependent attenuator (A) is used to represent the connector. The impedance (Z_{block}) and phase delay (θ_{block}) traversing the end block is represented by a transmission line. The impedance (Z_{sub}) and phase delay (θ_{sub}) traversing the RF substrate is represented by a second transmission line. The wire bond (L) connecting the fixture to the DUT is represented by an inductor.

To extract the model parameters (A, Z_{block}, θ_{block}, Z_{sub}, θ_{sub}, and L), de-embed fixtures should be made to provide short circuit and thru information. This is shown in Figure 9.3. The short circuit can be implemented with a via or solder joint to ground. An open circuit can be used in lieu of a short circuit, but with less accuracy. (The fringing fields at the end of the microstrip will be different in the test fixture than in the de-embed fixture).

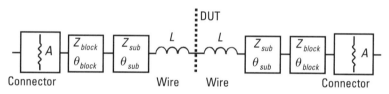

Figure 9.2   Equivalent circuit model for the test fixture shown in Figure 9.1.

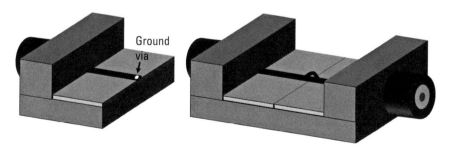

Figure 9.3   De-embed fixtures for the test fixture shown in Figure 9.1: short circuit (left) and thru circuit (right).

The shorted fixture (left image in Figure 9.3) is used to optimize S_{xx} magnitude and phase. The thru fixture (right image in Figure 9.3) is used to optimize S_{xy} magnitude and phase. Figure 9.4 shows this process.

In this two-port scenario, the values for A, Z_{block}, θ_{block}, Z_{sub}, and θ_{sub} can be optimized from the measured S_{11} magnitude and phase in the short-circuit fixture. Then, the value of L can be determined from the measured S_{21} magnitude and phase in the thru fixture. The process can be tailored for more than two-port circuits.

Once the equivalent circuit model has been determined, the *De_Embed-SnP* component in Keysight ADS can be used to mathematically de-embed the fixture behavior from the DUT performance.

9.2.2   Connectors, Adapters, and Cables

Generally, the test fixture is going to connect to a piece of measurement equipment (i.e., network or spectrum analyzer) through flexible cables. Benches are either set up for insertable or thru DUTs where the input and output connector sexes are different, or standard DUTs where the input and output sexes are the same (usually female).

All too often, connector, adapter, and cable selection is determined by parts available in the lab. Selection should be made based on frequency of operation, power-handling capability, and loss. These items all have maximum frequencies of operation where higher-order modes will begin to propagate. Flexible cables will also have a minimum bend radius specified by the manufacturer. Bending a cable at less than this angle can also excite higher-order modes. Higher-order modes can cause instability in active components or generate mixing products with the fundamental frequency, both invalidating the measurement.

There are scores of different connector, adapter, and cable types to meet a wide range of form factors, operating frequencies, and power-handling capabilities; the most popular types are described in the following subsections. As with most microwave components, the higher they operate in frequency, the more

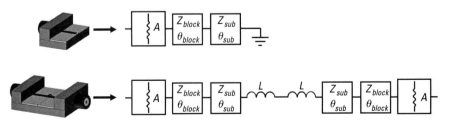

Figure 9.4   De-embed process using the fixtures shown in Figure 9.3 and the equivalent circuit model shown in Figure 9.2.

expensive they become. *Male* connectors typically have a center pin (*plug*) and internal threads, and *female* connectors have a center sleeve (*jack*) and external threads.

Bare wire

Typically used for DC connections, bare wires can support RF signals up to approximately 10 MHz [3]. To provide shielding, the signal (usually red) and ground (usually black) wires are twisted or braided together. When twisted tightly, they provide nearly the same electric shielding as a coax cable but with reduced magnetic shielding [3].

BNC

BNC connectors are generally used as a robust and easy-to-use (no torque wrench required) DC connector. They are relatively inexpensive and operate to 4 GHz. Due to their large size, they are better suited for test fixture usage than product usage.

N

N connectors are physically the largest listed in this text. They operate through X-band and some also operate through Ku-band. They are designed to handle high power and can easily operate with hundreds of watts and more than 1,000V root mean square (RMS). Since they are so rugged, connectors are often tightened by hand rather than using a torque wrench. For many applications, their size is prohibitively large.

Subminiature Version A (SMA)

SMA connectors are among the most popular in use today. The maximum operating frequency varies by manufacturer, but generally they are safe for use through Ku-band (higher-quality manufacturers will specify through K-band). Since they are PTFE- filled, SMA connectors tend to be more durable than air-filled varieties. Also, SMA connectors will mate with 3.5-mm and 2.92-mm connectors—although doing so is risky. The latter two are much more fragile (and expensive).

Practical Note

After spending weeks debugging a highly sophisticated measurement bench for a multimission radar emulator, the root cause was found to be a single bent center conductor on an SMA-to-2.92–mm adapter. The cost to replace the adapter was less than $5, while the cost to find the broken adapter was around $54,000. The moral: Always check connections carefully.

Subminiature Version B (SMB)

SMB connectors are smaller than SMA connectors. They operate from approximately DC to 4 GHz. Unlike SMAs, SMBs snap together rather than screw together.

3.5 mm

Probably the second most popular connector, 3.5-mm connectors are reasonably priced and operate to 26.5 GHz. Since they are air-filled, they are more fragile than SMAs.

2.92 mm (or simply 2.9 mm) or K

Similar in appearance to 3.5-mm connectors, 2.92-mm connectors have a smaller conductor diameter and operate to 40 GHz.

2.4 mm

While they resemble 2.92-mm connectors, 2.4-mm connectors have an even smaller conductor diameter. Due to their stringent manufacturing tolerances, 2.4-mm connectors are quite expensive, but they operate to 50 GHz.

Subminiature Push-On (SMP)

SMP connectors are useful when really small size is critical. They are especially popular for military applications. Similarly to SMBs, SMPs are also snap-on connections. However, SMPs are smaller than SMBs. There are several SMP sizes available that provide frequency coverage up to 65 GHz. Corning, Incorporated, has a line of SMP connectors that have been trademarked: GPO™, GPPO™, and G3PO™ (each one smaller and higher frequency than the prior).

Other Notable Connectors

- MCX: Similar to but smaller than SMB, MCX is a snap-on connector that operates to approximately 6 GHz;
- MMCX: The same style and operating frequency as the MCX, but smaller in size;
- TNC: Similar to BNC, but it operates through X-band and has a threaded connection;
- 1.85 mm: Smaller version of 2.4 mm that operates to 65 GHz;
- 1 mm: Smaller version of 1.85 mm that operates to 110 GHz.

9.3 Tips for Making it All Work

One always hopes that a design works exactly as expected the first time it is measured, but that isn't always the case. From the Monte Carlo analysis discussed in Section 8.6, the range of expected performance should be known. When measurements do not fall within that expected range, here are some tips for common issues and ways to approach them.

9.3.1 Unstable Active Circuits

When an active circuit is unstable, place a spectrum analyzer or oscilloscope on the DC bias or RF output to determine the frequency of oscillation. Then, if the oscillation is below the circuit operating frequency, try one or more of the following:

- Place a large-value capacitor between the drain and source voltages;
- Place a series inductor on the drain;
- Place a series resistor on the gate;
- Place a ferrite bead on the gate or drain bias;
- Verify that your DC bias wires are properly shielded. If bare wires are used, make sure that they are twisted or replace them with coax lines. Never cross gate bias wires and drain bias wires.

If the oscillation is within the circuit operating frequency, try one or more of the following:

- Add an attenuator to the input and/or output of the DUT to minimize VSWR reflection;
- Operate the circuit pulsed or if it is already operating pulsed, narrow the pulse width.

These are not necessarily long-term fixes, but once the DUT is stable, you can reanalyze and retune the circuit to improve stability.

9.3.2 Incorrect Frequency Response

If the frequency response has shifted or if signals appear at the output that are not supposed to be there, there are several things to investigate:

- Wires and ribbon bonds usually add 0.1–1-nH inductance everywhere they are placed. This can shift frequency, add phase length, and change the impedance. Ensure that all bonds are as small as possible (while still providing strain relief), and when possible, use multiple bonds to reduce inductance.

- Eliminate elbow bends or kinked cables from the measurement bench since they add parasitic inductance.

- Be mindful of background electrical noise sources, including the following (approximate frequencies):
  - 60-Hz and higher harmonics: Lights, power, and motors;
  - 535–1700 KHz: AM radio;
  - 30–300 MHz: FM radio and TV broadcasting;
  - 900 MHz: Cordless devices;
  - 2 GHz: Mobile phones;
  - 2.4 GHz or 5 GHz: Wireless LAN.

Intermittent ripple is usually caused by vibrations. Foot traffic, nearby elevators, or adjacent machinery can induce movement in bond wires or probe contact points. These small inductance changes can appear in the measured results.

Practical Note

It is not uncommon for labs near airports, train tracks, or major highways to conduct sensitive measurements only at night. Even when using an isolation table, periodic strong vibrations can affect measured results.

9.3.3 Radiation or Coupling

Radiation and coupling mechanisms are discussed in Section 8.3. To determine if airborne leakage is an issue—If safe to do so—move a piece of RF-absorbing material around the circuit while operational and look for changes in the measured response. Common problematic areas include the following:

- Metal sidewalls;
- Areas surrounding active circuits;
- Wraparound structures that connect top metal layers to bottom metal layers.

Do not place RF-absorbing material directly on top of transmission lines (which will only dampen the signal you are trying to protect) or sensitive structures (i.e., MMICs and bond wires). Once a source of radiation is found, shield the area properly or leave the absorbing material in place (if possible).

The same process can be repeated with aluminum foil or another conductive material. Be careful not to make contact with any conducting surface.

9.3.4　Low Gain or Output Power

If gain or output power is low, first correct any issues with stability and frequency response. If some frequencies in the operating band have much lower performance than others (referred to as *suck-outs*), check for evidence of radiation or coupling that could be canceling the signal at certain frequencies. If no issues are found, then consider the following:

- Measure the DC bias voltage as close to the circuit as possible. Sometimes a voltage drop occurs between the voltage source (i.e., DC supply) and the DUT. The circuit may be underbiased.

- Ensure that the DC bias current is as expected. Tweak the DC bias voltage to give the correct DC bias current (i.e., tweak the gate voltage to provide the correct quiescent current).

- Check the operating temperature. Gain and output power drop by approximately 0.016 dB/°C per stage so if the DUT is hotter than expected, that could explain the drop.

- (If safe to do so) step through the circuit with a pigtail from the input to the output to determine where in the circuit the performance deviates from the expected value. A pigtail is a short wire attached to an RF connector that can be attached to a network analysis, spectrum analyzer, power meter, oscilloscope, or similar measurement equipment. A well-made pigtail can operate to several gigahertz.

- Tune the circuit using ground pads or trombone structures as discussed in Section 7.6.2. Alternatively, cap sticks can be used to manually increase capacitance. Cap sticks can be made by gluing surface-mount capacitors to the end of a nonconductive stick. When pressed (or stacked) on top of another capacitor already in the circuit, the values add. Multiple cap sticks can be made with different values to tune the circuit. Once an optimized value is determined, the capacitor can be replaced with the higher value.

9.3.5   High Loss

In passive circuits, instability is not an issue, but excess loss can be an issue. Sources of loss can be determined in one of the following ways:

- Since even passive circuits radiate, verify that energy is not being lost by coupling to nearby circuits using the technique discussed in Section 9.3.3.

- Verify that line widths are as expected. This can usually be done using an optical microscope eyepiece reticle.

- Verify that the RF substrate thickness is as expected. This can usually be done using a blank substrate and a set of calipers. Remember to subtract the top and bottom metal thickness.

- Since transmission lines with excessive surface roughness will have higher than expected metal loss, assess roughness by comparing the metal shine against a known-good sample. Rough surfaces appear less shiny than smooth surfaces. If available, roughness can also be measured using a surface profilometer.

- Check surface-mount components to ensure that they are well attached and are of the correct value.

Practical Note

Digital RLC (or LCR) tweezers are useful for measuring surface-mount resistors, inductors, and capacitors. They can be used to quickly verify that all components within a circuit are the correct value.

9.3.6   Catastrophic Damage at Initial Test

The most difficult type of failure to diagnose is when catastrophic damage (i.e., a transistor pops) occurs at initial test. Even a DUT with poor performance provides hardware that can usually be investigated. Catastrophic damage at initial test can usually be avoided with the following precautions:

- For active circuit, apply DC bias in stages.

- Pinch off the gate with a very small drain voltage applied; verify that no drain current is present.

- Increase the drain voltage to the operating value; verify that no drain current is still present.

- Slowly bring the gate out of pinch-off and watch for a steady increase in the drain current; stop if any fluctuations are found in the drain current as this is a sign of instability.
- Once the quiescent bias is met, apply the RF signal.
- Verify that all cables are placed within their rated bend radius. If this radius is exceeded, the cable dielectric separates from the conductor and a charge builds up (known as the *triboelectric effect*). This charge can dissipate at any time and cause failures to occur.

9.4 Transistor Stabilization

When a radar is deployed, it needs to provide consistent behavior over the operating lifetime. At the component level, an additional process step must be performed whenever active circuits are used. When a transistor is operated for the very first time, output power, gain, and drain current drop and gate current rises over a relatively short period of time (usually minutes or hours). Fortunately, if operated long enough, the performance will reach an equilibrium point or stabilize (not be confused with electrical stability, discussed in Chapter 4). Once a transistor has stabilized, the power, gain, and current will remain consistent throughout the operational lifetime. Figure 9.5 shows a typical set of stabilization curves for a fixed-input power and bias level; all values are normalized.

Usually, output power settles first and happens quickly in the stabilization process. A good rule of thumb is the 90/10 rule; 90% of output power stabilization happens in the first 10% of the process. Drain current usually settles second and follows an 80/20 rule. Gate current settles last and ultimately defines

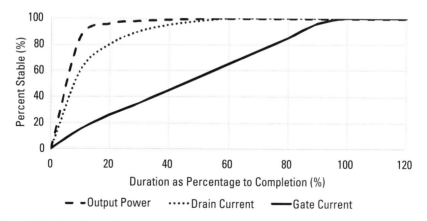

Figure 9.5 Typical set of normalized transistor stabilization curves.

the stabilization duration. Once all three levels are settled, they will remain flat even if the component is turned off and rebiased at a later time. Every foundry process is different so not all transistors follow this exact process.

The duration required to achieve stabilization depends on temperature; the higher the temperature, the shorter the required stabilization duration. However, the operating lifetime of a transistor is also dependent on temperature; the higher the temperature, the shorter the operating lifetime. The temperature must be chosen so that stabilization can be achieved in a reasonable amount of time but does not take useable life out of the transistor.

Usually, foundries can provide a suggested stabilization process (temperature, D/C bias, input power level, and duration). From this starting point, the duration can be customized using the Arrhenius equation, (9.1) [4].

$$\ln\left(\frac{\tau_1}{\tau_2}\right) = \frac{E_a}{k}\left(\frac{1}{T_2} - \frac{1}{T_1}\right)$$

(9.1)

where τ_N is the duration at T_N (any time unit), T_N is the temperature (K), E_a is the activation energy (electronvolts, eV), and k is Boltzmann's constant (8.6 $\times 10^5$ eV/K).

For example, if a transistor has an activation energy of 1.5 eV and a stabilization process of 100 hours at 85°C (358K) base temperature, the same stabilization could be achieved in 14 hours at 100°C (373K) base temperature (provided that the operational lifetime is still sufficient after operating at 100°C base temperature).

Practical Note

The physical location of the base temperature can vary from foundry to foundry. Be sure to differentiate between the channel or junction, the backside of the transistor, and the backside of the carrier if the transistor is packaged. The channel or junction temperature will be hotter than the surrounding area.

Not all applications require stable performance. If the radar system has the ability to adjust input power and DC bias, then having transistors that change performance over time is not an issue. For these applications, a *screening* process is needed to expose components with latent defects (issues that cannot be found with normal inspection). Generally the requirements for screening are determined empirically. A large sample set of components is tested to determine how long it takes for a typical defected unit to fail. That duration is multiplied by a safety factor to provide margin and becomes the screening criteria.

Exercises

1. A noise measurement is unable to differentiate between the noise generated from the DUT and background noise. What can be done to improve this measurement?

2. You are designing an amplifier module that must have at least 100-W (50 dBm) output power. The measurement uncertainty for the test bench is ±0.5 dB. In order to compensate for this uncertainty, what should your design goal be (assume that margin is already factored into the 100-W specification)?

3. It has been shown that ribbons need to be as short as possible to minimize inductance, but long enough to provide strain relief. How much height is needed for a 10-mil-wide, 1-mil-thick gold ribbon that traverses a 10-mil-wide gap? Assume that the storage temperature ranges from −50°C to +150°C.

4. When searching for airborne coupling, does the technique vary depending on the frequency of operation? Why or why not?

5. Bench-level circuit tuning can be performed using cap sticks as described in Section 9.3.4. Could the same be done with inductor sticks or resistor sticks? When might this be a bad idea?

6. You are operating a radar submodule at 1.5 GHz. On the spectrum analyzer, you also see a tone at 900 MHz. How can you determine if the tone is from an outside source or from the module itself?

7. A transistor with a bandgap of 2 eV requires 24 hours to stabilize at 100°C.

   • How long would it take to stabilize at 110°C?

   • How long would it take to stabilize at 90°C?

References

[1] Agilent Technologies, "Agilent Fundamentals of RF and Microwave Power Measurements (Part 3): Power Measurement Uncertainty per International Guides," Internet, http://cp.literature.agilent.com/litweb/pdf/5988-9215EN.pdf.

[2] Wartenberg, S., "RF Test Fixture Basics," *Microwave Journal,* June 2003.

[3] Ott, H., *Noise Reduction Techniques in Electronic Systems,* New York, NY: John Wiley & Sons, Inc., 1988.

[4] Cheney, D., "Degradation Mechanisms for GaN and GaAs High Speed Transistors," *Materials,* No. 5, 2012, pp. 2498–2520.

Selected Bibliography

Laverghetta, T., *Modern Microwave Measurements and Techniques,* Norwood, MA: Artech House, 1988.

Schaub, K., and J. Kelly, *Production Testing of RF and System-on-a-Chip Devices for Wireless Communications,* Norwood, MA: Artech House, 2003.

Wartenberg, S., *RF Measurements of Die and Packages,* Norwood, MA: Artech House, 2001.

10

Final Thoughts

The proliferation of radars in the commercial and military markets is going to continue for decades to come. This book is intended as a reference for engineers embarking on the challenging and rewarding path of component design for radars and a guidebook for how that design impacts overall system performance and cost. As the book comes to a close, please consider these top takeaway points, summarized as follows.

- Collaboration between system and component designers before hardware is built is critically important to truly optimize performance. The best way to facilitate this dialog is by understanding both sides. This book provides the background necessary for system engineers to design components and for component engineers to optimize for radar needs.

- Allow for design iterations, especially between system designers and components engineers. It's impossible to get it "perfect" on the first pass! Many major programs have incurred significant schedule and cost overruns by not embracing this design reality.

- Since understanding design equations is beneficial for building an intuition of trade-offs, review them carefully and remember the variables that contribute to critical performance metrics, such as power, efficiency/loss, size, and heat removal. The exercises provided at the end of each chapter are an excellent way to both test your knowledge and grow your intuition.

- Leverage design software, such as Keysight ADS, to expedite the design process. Use the simulator to determine integrated performance under all operating conditions. (Don't forget to include bond wires and other connection mechanisms.)

- Choose the right method for mode propagation (i.e., coax, microstrip, stripline, waveguide, or coplanar waveguide) based on loss, size, operating frequency, dispersion, cost, and producibility.

- Pay close attention to material selection. Make the decision based on physical, electrical, mechanical, thermal, chemical, manufacturing, and environmental properties.

- Take great care to ensure that component models are accurate. Try to capture as many parasitic or nonideal effects as possible. Mitigate discontinuities along the signal path.

- When designing amplifiers, expand beyond the ubiquitous class AB. There are many other techniques that offer better efficiency, linearity, output power, and operating bandwidth.

- Although noise in amplifier components is unavoidable, minimize it with proper material selection, DC biasing, impedance, and temperature management.

- Design components with both electrical and mechanical performance in mind. Having phenomenal electrical performance is meaningless if the component degrades due to thermal or structural stress.

- Give passive components the same level of attention as every other component. Active components usually get the most attention, but passive components are often what drive size, cost, and loss.

- Remember that determining the ideal form factor is an important decision. MMICs, hybrids, multichip modules, and the various packaging styles all have unique attributes that contribute to size, cost, and performance.

- Design with manufacturing in mind from the beginning of the process. Cost, yield, and lifetime can all be significantly improved by considering these metrics during the design phase.

- Maintain proper grounding so that signals propagate as intended. The old adage, "you can't have enough ground connections," is not necessarily true; you need the right kind of ground connections.

- Use shielding, orientation, and spacing to isolate components, since radiation or coupling effects between components can degrade performance or even cause failure.

- Prevent the many types of oscillations possible (i.e., even-mode, odd-mode, spurious, and loop) with good planning, layout, and simulation. These can be tricky to diagnose on the bench.

- Use Monte Carlo analysis to predict and mitigate variability of performance during manufacturing. Any and all contributors to performance variation should be included for the analysis to be accurate.

- Integrate digital and RF circuitry to enable a new level of performance potential, but do it carefully. Otherwise, the circuitries' proximity to each other can cause performance degradation and other issues.

- Know the accuracy and precision of the measurement bench. Some degree of uncertainty is present in every test system. That uncertainty should be factored into every performance specification.

- Choose the right connector for your application. Many options are available based on size, cost, loss, power handling, durability, and frequency of operation. Make sure the one you select does not sacrifice.

- Remember that while all components benefit from screening for latent defects, some semiconductors require an additional stabilization process. Ignoring this step could result in a performance drop during operation.

Appendix A

Tables A.1–A.10, with universal constants and material properties, are provided to serve as a design reference. The material properties listed are intended only as guidelines. Since properties vary, always use the performance provided by the supplier. Where appropriate, a range of values is provided to show the variation of performance across suppliers.

A.1 Frequency Bands

Table A.1

Frequency Bands

Letter Designation	Frequency Range (gigahertz)	Free-Space Wavelength (millimeters)
HF	0.003–0.03	100,000–10,000
VHF	0.03–0.3	10,000–1,000
UHF	0.3–1	1,000–300
L	1–2	300–150
S	2–4	150–75
C	4–8	75–37.5
X	8–12	37.5–25
Ku	12–18	25–16.7
K	18–26.5	16.7–11.3
Ka	26.5–40	11.3–7.5
V	40–75	7.5–4
W	75–110	4–2.7
Millimeter-wave	30–300	10–1
Terahertz	300–3,000	1–0.1

A.2   English-to-Metric Units Conversion

Table A.2
Metric Headers

Metric Header	Symbol	Decimal	Written
Tera	T	10^{12}	Trillion
Giga	G	10^{9}	Billion
Mega	M	10^{6}	Million
Kilo	K	10^{3}	Thousand
Unit	—	1	One
Centi	c	10^{-2}	Hundredth
Milli	m	10^{-3}	Thousandth
Micro	μ	10^{-6}	Millionth
Nano	n	10^{-9}	Billionth
Pico	p	10^{-12}	Trillionth
Femto	f	10^{-15}	Quadrillionth

Table A.3
Metric-to-English Unit Conversion

	Metric	English
Mass	1g	0.03527 oz
	1 oz	28.3495g
	1 kg	2.2046 lbs
	1 lbs	0.4536 kg
Length	1 cm	0.3937 in
	1 in	2.54 cm
	1m	3.2808 ft
	1 ft	0.3048m
	1 ft	12 in
	1 mm	39.37 mils
	1 mils	25.4 μm
Volume	1L	61.0237 in^3
	1 ft^3	28.3168L

A.3 Temperature Conversion

$$T(^\circ C) = \frac{5}{9}\left[T(^\circ F) - 32\right] = T(K) - 273.15 \qquad (A.1)$$

$$T(^\circ F) = \frac{9}{5}T(^\circ C) + 32 = \frac{9}{5}T(K) - 459.67 \qquad (A.2)$$

$$T(K) = \frac{5}{9}\left[T(^\circ F) - 32\right] + 273.15 = T(^\circ C) + 273.15 \qquad (A.3)$$

A.4 Constants and Material Properties

Table A.4
Universal Constants

Constant	Symbol	Value
Base of natural logarithms	e	2.71828183
Boltzmann's constant	k	$1.3806488 \times 10^{-23}$ J/K 8.6173324×10^{-5} eV/K
Electron mass	m_e	$9.10938291 \times 10^{-31}$ kg
Free-space permeability	μ_o	$4\pi \times 10^{-7}$ H/m $\approx 1.25663706144 \times 10^{-6}$ H/m
Free-space permittivity	ε_o	$\dfrac{1}{c^2 \mu_o}$ F/m $\approx 8.85418781762 \times 10^{-12}$ F/m
Fundamental charge	e	$1.60217665 \times 10^{-19}$ C
Pi	π	3.1415926536
Speed of light in free space	c	2.99792458×10^8 m/s
Stefan-Boltzmann constant	σ_{SB}	5.670373×10^{-8} W/m²·K⁴

Table A.5

RF Substrate Properties

Material	Dielectric Constant (ε_r)	Loss Tangent ($\tan \delta$)
Air	1.0005	Weather-dependent
Aluminum nitride	8.8	0.001
Aluminum oxide	8.8–10.1	0.0002–0.002
Barium titanate	1,200	0.013
Beryllium Oxide	6.4–6.7	0.0003–0.003
Gallium arsenide	12.88–12.9	0.0004–0.001
HTCC*	9.5	0.0004
Indium phosphide	12.4	0.006
Liquid crystal polymer	2.9	0.002
LTCC†	5	0.0002
PTFE (Teflon) ‡	2.08–2.84	0.00015–0.002
Silicon	11.68–12.9	0.00075–0.003
Silicon carbide	10.8	0.003
Silicon dioxide	3.8–3.9	0.0008
Silicon nitride	7.5	0.003–0.006
Titanium dioxide	86–173	0.0015–0.002
Vacuum	1	0

* High-temperature cofired ceramic
† Low-temperature cofired ceramic
‡ Polytetrafluoroethylene

Table A.6

Metal Electrical Conductivity

Material	Electrical Conductivity (σ, S/m)
Aluminum	3.5×10^7–3.82×10^7
Brass	1.5×10^7–1.59×10^7
Copper	5.80×10^7–5.96×10^7
Gold	4.10×10^7
Nichrome	6.7×10^5–9.09×10^5
Nickel	1.43×10^7–1.45×10^7
Platinum	9.43×10^6
Silver	6.17×10^7–6.30×10^7
Tin	9.17×10^6
Titanium	2.38×10^6
Tungsten	1.79×10^7–1.82×10^7

Table A.7

Semiconductor Bandgap

Material	Bandgap (eV)
Carbon (diamond)	5.4–5.47
Gallium arsenide	1.35–1.43
Gallium nitride	3.44
Indium phosphide	1.27–1.35
Silicon	1.107–1.12
Silicon carbide	2.3–3.3

Table A.8

Mechanical Properties: Young's Modulus and Poisson's Ratio

Material	Young's Modulus, E (GPa)	Poisson's Ratio (v)
Aluminum	89–79	0.33–0.35
Brass	97–110	0.34
Bronze	97–117	0.34
Copper	110–124	0.33–0.36
Gallium arsenide	85.5	0.31
Gold	79	0.4
Kovar	138	0.317
Nickel	207	0.31
Silicon	129.5–186.5	0.22–0.28
Titanium	103–117	0.33
Tungsten	345–379	0.2

Table A.9

Mechanical Properties: Coefficient of Thermal Expansion, Thermal Conductivity,
Specific Heat, and Density

Material	Coefficient Thermal Expansion, α (10⁻⁶/°C)	Thermal Conductivity, k (W/cm–K)	Specific Heat, c (J/g·°C)	Density, ρ (kg/m³)
Air	—	263	1.02	1.1614
Aluminum	23–25	2.35	0.9	2700
Aluminum nitride	4.0–5.3	1.6–2.3	0.74	3260
Aluminum oxide	6.5–8.4	0.0035–0.0037	0.88	3800
Beryllium oxide	6.9–9.0	2.00–3.00	1.02–1.12	3000
Brass	19.1–21.2	1.2	0.38	8430–8730
Bronze	18–21	1.1	0.435	7400–8920
Copper	16.6–17.6	4.01	0.385	8960
Gallium arsenide	5.39–6.86	0.46	0.35	5320
Gold	14–14.2	3.45	0.131	19320
Gold tin solder	16	0.57	0.15	14700
Helium	—	1,520	5.3	0.1625
Hydrogen	—	1,830	14.267	0.0808
Kovar	4.9–6.2	0.167–0.17	0.46	8000–8400
LTCC	3.0–5.8	0.002–0.003	0.989	2600
Nickel	13	1.58	0.444	8908
Silicon	2.56–2.6	1.3–1.48	0.7–0.712	2329
Silicon carbide	2.77–4.0	4.9	0.75	3160
Silver	18	4.28	0.234	10490
Silver epoxy	54–200	0.008–0.02	0.787	10490
Tin	23.4	0.85	0.226	7280
Titanium	8.1–11	0.31	0.523	4500
Tungsten	4.3–4.5	1.73–2.35	0.134	19450
Water	69	0.0058	4.184	1003

Table A.10
Material Emissivity (All Metals
Are Polished)

Material	Emissivity, ε
Aluminum	0.04–0.05
Aluminum (foil)	0.07–0.09
Aluminum oxide	0.69
Brass	0.03
Chromium	0.1
Copper	0.03–0.05
Gold	0.02–0.03
Gold (foil)	0.07–0.09
Nickel	0.05
Silicon carbide	0.83–0.96
Silver	0.02–0.03

A.5 Math Functions

To serve as a design reference, the following math functions are provided.

Log Rules

Common log: $10^x = y$ $\log y = x$ $10^{\log A} = A$ $\log 10^A = A$

Natural log: $e^x = y$ $\ln y = x$ $e^{\ln A} = A$ $\ln e^A = A$

$\log 1 = \ln 1 = 0$ $\log 10 = 1$ $\ln e = 1$

$\ln A = (\ln 10)\log A \approx 2.3026 \log A$ $\log A = (\log e)\ln A \approx 0.4342 \ln A$

$\log AB = \log A + \log B$ $\log \dfrac{A}{B} = \log A - \log B$ $\log \dfrac{1}{A} = -\log A$

$\log A^n = n \log A$

Exponent Rules

$A^n A^m = A^{n+m}$ $(A^m)^n = A^{mn}$ $(AB)^m = A^m B^m$

$\left(\dfrac{A}{B}\right)^m = \dfrac{A^m}{B^m}$ $A^{\frac{m}{n}} = \sqrt[n]{A^m}$ $A^0 = 1 \; (A \neq 0)$

About the Authors

Nickolas Kingsley has spent more than 15 years designing microwave front-end components for commercial and military applications. He is an avid designer and inventor and has given invited talks internationally on topics ranging from device physics to next-generation system architectures. He was with Auriga Microwave for eight years and during his tenure served as director of engineering. He currently leads the RF front-end and converter capability group at BAE Systems in Nashua, New Hampshire.

Dr. Kingsley has a Ph.D. in electrical engineering from the Georgia Institute of Technology. His specializations include active and passive design, packaging, multiphysics analysis, and designing for manufacturing. He is a senior member of the IEEE, a technical program review committee chair for the International Microwave Symposium, an active member of the Microwave Theory and Techniques Society, and a technical reviewer for nearly a dozen publications.

Joseph R. Guerci has over 30 years of advanced technology development experience in industrial, academic, and government settings—the latter included a seven-year term with Defense Advanced Research Projects Agency (DARPA) where he led major new technology development efforts in his successive roles as program manager, deputy office director, and director of the special projects office. He is currently president and CEO of Information Systems Laboratories, Inc.

Dr. Guerci, who has a Ph.D. in electrical engineering from NYU Polytechnic University, is the author of over 100 technical papers and publications, including the books *Space-Time Adaptive Processing for Radar,* 2nd ed., (Artech

House) and *Cognitive Radar: The Knowledge-Aided Fully Adaptive Approach* (Artech House). He is a fellow of the IEEE for *Contributions to Advanced Radar Theory and its Embodiment in Real-World Systems,* and the recipient of the 2007 IEEE Warren D. White Award for *Excellence in Radar Adaptive Processing and Waveform Diversity.*

The views expressed in this book are those of the authors and do not reflect the views, policy, or position of BAE Systems.

Index

291

Recent Titles in the Artech House Radar Series

Dr. Joseph R. Guerci, Series Editor

Electronic Warfare Target Location Methods, Second Edition,
  Richard A. Poisel

ELINT: The Interception and Analysis of Radar Signals,
  Richard G. Wiley

EW 101: A First Course in Electronic Warfare, David Adamy

EW 102: A Second Course in Electronic Warfare, David Adamy

EW 103: Tactical Battlefield Communications Electronic Warfare,
  David Adamy

Fourier Transforms in Radar and Signal Processing, Second Edition,
  David Brandwood

Fundamentals of Electronic Warfare, Sergei A. Vakin, Lev N. Shustov,
  and Robert H. Dunwell

Fundamentals of Short-Range FM Radar, Igor V. Komarov and
  Sergey M. Smolskiy

*Handbook of Computer Simulation in Radio Engineering,
  Communications, and Radar,* Sergey A. Leonov and
  Alexander I. Leonov

High-Resolution Radar, Second Edition, Donald R. Wehner

Highly Integrated Low-Power Radars, Sergio Saponara, Maria Greco,
  Egidio Ragonese, Giuseppe Palmisano, and Bruno Neri

Introduction to Electronic Defense Systems, Second Edition,
  Filippo Neri

Introduction to Electronic Warfare, D. Curtis Schleher

Introduction to Electronic Warfare Modeling and Simulation,
  David L. Adamy

Introduction to RF Equipment and System Design, Pekka Eskelinen

Introduction to Modern EW Systems, Andrea De Martino

The Micro-Doppler Effect in Radar, Victor C. Chen

Microwave Radar: Imaging and Advanced Concepts,
  Roger J. Sullivan

Millimeter-Wave Radar Targets and Clutter, Gennadiy P. Kulemin

Radar Technology Encyclopedia, David K. Barton and
   Sergey A. Leonov, editors

Radio Wave Propagation Fundamentals, Artem Saakian

Range-Doppler Radar Imaging and Motion Compensation,
   Jae Sok Son, et al.

Robotic Navigation and Mapping with Radar, Martin Adams,
   John Mullane, Ebi Jose, and Ba-Ngu Vo

Signal Detection and Estimation, Second Edition, Mourad Barkat

Signal Processing in Noise Waveform Radar, Krzysztof Kulpa

Space-Time Adaptive Processing for Radar, Second Edition,
   Joseph R. Guerci

Special Design Topics in Digital Wideband Receivers, James Tsui

Theory and Practice of Radar Target Identification,
   August W. Rihaczek and Stephen J. Hershkowitz

Time-Frequency Signal Analysis with Applications, Ljubiša Stanković,
   Miloš Daković, and Thayananthan Thayaparan

*Time-Frequency Transf   orms for Radar Imaging and Signal
   Analysis*, Victor C. Chen and Hao Ling

Transmit Receive Modules for Radar and Communication Systems,
   Rick Sturdivant and Mike Harris

For further information on these and other Artech House titles, including previously considered out-of-print books now available through our
In-Print-Forever® (IPF®) program, contact:

Artech House	Artech House
685 Canton Street	16 Sussex Street
Norwood, MA 02062	London SW1V HRW UK
Phone: 781-769-9750	Phone: +44 (0)20 7596-8750
Fax: 781-769-6334	Fax: +44 (0)20 7630-0166
e-mail: artech@artechhouse.com	e-mail: artech-uk@artechhouse.com

Find us on the World Wide Web at: www.artechhouse.com